Letts

# Revise GCSE

# Mathematics

Gillian Rich

# Contents

# Your GCSE course

## Use these pages to get to know your course

- Make sure you know your exam board
- Check which specification you are doing

- Know how your course is assessed:
  - What format are the papers?
  - How many papers are there?

## Mathematics

| Exam board | Syllabus number | Availability | Assessment |
|---|---|---|---|
| AQA | 4360 (Unitised) | November March June | Statistics and Number (calculator) Unit 1F: 1hr, 54 marks, 26.7% Unit 1H: 1hr, 54 marks, 26.7% Number and Algebra (non-calculator) Unit 2F: 1hr 15mins, 66 marks, 33.3% Unit 2H: 1hr 15mins, 66 marks, 33.3% Geometry and Algebra (calculator) Unit 3F: 1hr 30mins, 80 marks, 40% Unit 3H: 1hr 30mins, 80 marks, 40% |
| AQA | 4365 (Linear) | January June | Foundation: Paper 1F (non-calculator): 1hr 15mins, 70 marks, 40% Paper 2F (calculator): 1hr 45mins, 105 marks, 60% Higher: Paper 1H (non-calculator): 1hr 30mins, 70 marks, 40% Paper 2H (calculator): 2hrs, 105 marks, 60% |
| EDEXCEL A | Linear 1MA0 | March June November | Foundation: Paper 1F (non-calculator): 1hr 45mins, 100 marks, 50% Paper 2F (calculator): 1hr 45mins, 100 marks, 50% Higher: Paper 1H (non-calculator): 1hr 45mins, 100 marks, 50% Paper 2H (calculator): 1hr 45mins, 100 marks, 50% |
| EDEXCEL B | Modular 2MB01 | March June November | Statistics and Probability (calculator) Unit 1F: 1hr 15mins, 60 marks, 30% Unit 1H: 1hr 15mins, 60 marks, 30% Number, Algebra and Geometry 1 (non-calculator) Unit 2F: 1hr 15mins, 60 marks, 30% Unit 2H: 1hr 15mins, 60 marks, 30% Number, Algebra and Geometry 2 (calculator) Unit 3F: 1hr 30mins, 80 marks, 40% Unit 3H: 1hr 45mins, 80 marks, 40% |

# Mathematics

| Exam board | Syllabus number | Availability | Assessment |
|---|---|---|---|
| OCR A | J562 | January<br>June<br>November | Unit A (calculator)<br>Paper 1F: 1hr, 60 marks, 25%<br>Paper 1H: 1hr, 60 marks, 25%<br><br>Unit B (non-calculator)<br>Paper 2F: 1hr, 60 marks, 25%<br>Paper 2H: 1hr, 60 marks, 25%<br><br>Unit C (calculator)<br>Paper 3F: 1hr 30mins, 100 marks, 50%<br>Paper 3H: 2hrs, 100 marks, 50% |
| OCR B | J567 | June | Foundation:<br>Paper 1F (non-calculator): 1hr 30mins, 100 marks, 50%<br>Paper 2F (calculator): 1hr 30mins, 100 marks, 50%<br><br>Higher:<br>Paper 3H (non-calculator): 1hr 45mins, 100 marks, 50%<br>Paper 4H (calculator): 1hr 45mins, 100 marks, 50% |
| WJEC | Unitised | November<br>January<br>June | Unit 1: Mathematics in Everyday Life (calculator)<br>Paper F: 1hr 15mins, 65 marks, 30%<br>Paper H: 1hr 15mins, 65 marks, 30%<br><br>Unit 2: Non-calculator Mathematics<br>Paper F: 1hr 15mins, 65 marks, 30%<br>Paper H: 1hr 15mins, 65 marks, 30%<br><br>Unit 3: Calculator-allowed Mathematics<br>Paper F: 1hr 30mins, 80 marks, 40%<br>Paper H: 1hr 45mins, 90 marks, 40% |
| WJEC | Linear | June<br>November | Foundation:<br>Paper 1 (non-calculator): 1hr 45mins, 100 marks, 50%<br>Paper 2 (calculator): 1hr 45mins, 100 marks, 50%<br><br>Higher:<br>Paper 1 (non-calculator): 2hrs, 100 marks, 50%<br>Paper 2 (calculator): 2hrs, 100 marks, 50% |
| CCEA | 2210 | January<br>June | Foundation:<br>Unit T1 (calculator) 1hr 30mins, 100 marks, 45%<br>or<br>Unit T2 (calculator) 1hr 30mins, 100 marks, 45%<br>and<br>Unit T5: Paper 1 (non-calculator) 1hr, 50 marks,<br>and Paper 2 (calculator) 1hr, 50 marks, 55%<br><br>Higher:<br>Unit T3 (calculator) 2hrs, 100 marks, 45%<br>or<br>Unit T4 (calculator) 2hrs, 100 marks, 45%<br>and<br>Unit T6: Paper 1 (non-calculator) 1hr 15mins, 50 marks,<br>and Paper 2 (calculator) 1hr 15mins, 50 marks, 55% |

# How to use this book

For all specifications the available grades are the same. Contact your awarding body for full details of your specifications.

| Tier | Grades available |
|---|---|
| Foundation | C–G |
| Higher | A*–D (E allowed) |

www.aqa.org.uk
www.ocr.org.uk
www.edexcel.com
www.wjec.co.uk
www.rewardinglearning.org.uk

## What this book covers

- *Revise GCSE Mathematics* provides full coverage of all the content and skills on the examination board specifications. It is divided into six chapters:
  Chapter 1: Number
  Chapter 2: Algebra
  Chapter 3: Geometry
  Chapter 4: Measures
  Chapter 5: Statistics
  Chapter 6: Probability
- The content is accessible to students working at both Foundation and Higher Tier levels.
- The practice questions reflect the level of questions on the Higher Tier exam papers. They are intended to be challenging to help you push yourself to achieve the best possible grade.

## Functional skills

Functional skills are skills that help you in everyday and real life situations. They are an important part of the new GCSE Mathematics courses. This book includes questions that require you to apply your knowledge of maths to solve problems relevant to real life.

## How this book will help you

- The summary table and specification labels will give you a quick reference to the requirements for your examination.

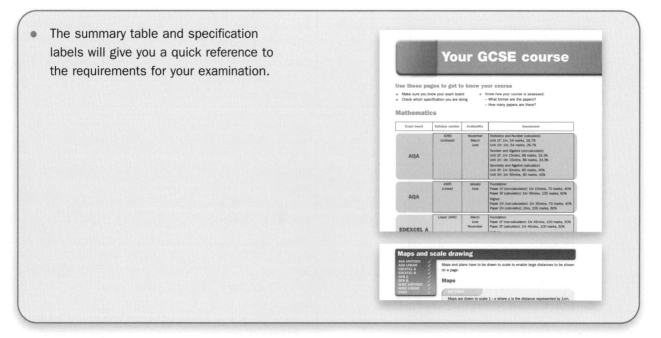

- Key points highlight important revision facts and methods.

- Important mathematical terms are highlighted in bold. Make sure you understand what these terms mean.

- Useful hints in the margin give guidance and focus your attention on important revision tips.

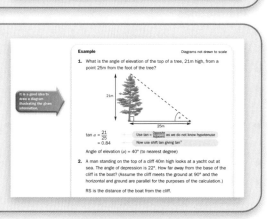

- Worked examples take you through calculations and problem solving one step at a time.

- Progress checks and exam practice questions help you to confirm your understanding of the topic.

- You will need to do your working and write your answers on a separate piece of paper. Check your answers against those in the book. If some are incorrect, then go back over the topic to see where you went wrong.

- (calculator icon) This icon indicates that the question should be attempted without a calculator.

- Sample GCSE questions are given with model answers and advice on how to gain the most marks.

  *N.B. Do not worry if the questions in your exam paper do not follow exactly the same style as you will find in this book. There are a variety of question styles provided here to give you the best practice possible.*

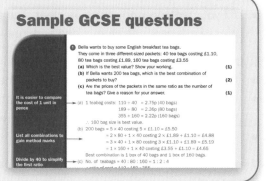

# Five ways to improve your grade

## 1. Preparation

- Make a list of **revision topics** and leave plenty of time to revise.
- Check that all your **maths equipment** is ready for the exam. You need a pen, pencil, ruler, rubber, pencil sharpener, protractor, pair of compasses and calculator.
- Make sure you know how to use your **calculator** efficiently – read the instruction book!
- Learn the different methods of working through a question.
- **Practise** answering questions under timed conditions.

## 2. Instructions

- Make sure you know how long the exam lasts and the end time. You should then be able to **pace yourself** through the paper.
- **Read carefully** the instructions on the front of the exam paper.
- Some **formulae** are printed inside the cover of the exam paper. Remember that they are there – you may need to use them.
- You must use a pen to write on the exam paper, but make sure you use a pencil for graphs and diagrams; if you make a mistake it is easier to correct.
- Write all working and answers in the spaces provided.
- You must answer all questions to obtain full marks. At least make an attempt at all questions because it is better than leaving a question out completely.

## 3. Questions

- **Read every question carefully**, so you know what is required.
- Do not spend too much time on a question. You can come back to it later.
- Marks allocated and space given are an indication of the length of answer required.
- Questions with more marks are usually divided into smaller steps. Work through the question from the first step.
- **Method marks** are given for each step even if the final answer is incorrect.
- Even **rough working** must be written on your exam paper – it is the only paper you will have.
- Do not do your rough working on your hand, pencil case or any other surface. You cannot give it in, so the examiner will not see it and you may lose marks.

## 4. Working

- **Graphs** must be drawn with a sharp pencil, never pen. Remember to label axes.
- Always give **units** in the answers if they are not provided.
- Answers must be given to an **appropriate degree of accuracy**.
- Write down the **calculator display** in full. Leave **rounding and correction** until the final answer, unless directed otherwise.
- Marks are given for **method, facts and answers**.
- If an incorrect answer is given without working, the mark will be zero.
- All your **written work must be clear**. The examiner is not used to your handwriting. Untidy, confused and messy answers can be misread and lose marks.

## 5. Checking

- Check that all answers are sensible in the context of the question.
- Check that you have answered or attempted every question. Have you missed anything? It would be a shame to lose marks because you have turned over two pages at once!
- Check through your paper for at least 10 minutes before the examination ends.

**If you have prepared and revised well, you can approach the examination calmly and with confidence. Good luck!**

# 1 Number

The following topics are covered in this chapter:

- Integers
- Powers and roots
- Fractions
- Decimals
- Percentages
- Ratio and proportion
- Approximations
- Calculator use

# 1.1 Integers

**LEARNING SUMMARY**

After studying this section, you should be able to understand:

- order of operations
- positive and negative integers
- common factors and multiples
- prime numbers

## Order of operations

AQA UNITISED ✓
AQA LINEAR ✓
EDEXCEL A ✓
EDEXCEL B ✓
OCR A ✓
OCR B ✓
WJEC UNITISED ✓
WJEC LINEAR ✓
CCEA ✓

**KEY POINT**

Remember the order of working by using BIDMAS:

**B**rackets **I**ndices **D**ivision **M**ultiplication **A**ddition **S**ubtraction

(Sometimes you will see BODMAS. The O stands for orders, i.e. powers.)

**Example**

Work out $6^2 - 3(2 + 5)$

$$6^2 - 3(2 + 5) = 6^2 - 3(7)$$
$$= 6^2 - 21$$
$$= 36 - 21$$
$$= 15$$

## Positive and negative integers

AQA UNITISED ✓
AQA LINEAR ✓
EDEXCEL A ✓
EDEXCEL B ✓
OCR A ✓
OCR B ✓
WJEC UNITISED ✓
WJEC LINEAR ✓
CCEA ✓

**Integers** are whole numbers.

- Positive numbers are greater than zero (+).
- Negative numbers are less than zero (-).

## Adding and subtracting integers

Integers can be shown on a number line.

-6  -5  -4  -3  -2  -1  0  1  2  3  4  5  6

To add move right →
To subtract move left ←

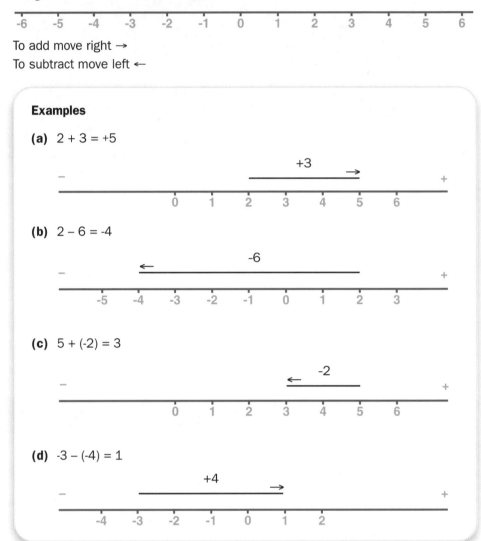

**Examples**

**(a)** 2 + 3 = +5

**(b)** 2 – 6 = -4

**(c)** 5 + (-2) = 3

**(d)** -3 – (-4) = 1

Adding a negative number is the same as subtracting.
If there is no sign, the integer is positive.

Subtracting a negative number is the same as adding.

**KEY POINT**

When adding and subtracting integers, these rules are used:
Same signs (+ +) or (− −) = +
Different signs (+ −) or (− +) = −

For example:
- 3 + (-6) = 3 − 6 = -3
- 50 − (-5) = 50 + 5 = +55

## Multiplying and dividing integers

Use the (+/−) or (−) calculator key to change the sign.

The rules for adding and subtracting integers also apply to multiplying and dividing, i.e. two like signs give a positive answer and two different signs give a negative answer.

For example:
- (-3) × (+6) = -18
- (-50) ÷ (-5) = +10

# Common factors and multiples

| | |
|---|---|
| AQA A | ✓ |
| AQA B | ✓ |
| EDEXCEL A | ✓ |
| EDEXCEL B | ✓ |
| OCR A | ✓ |
| OCR B | ✓ |
| WJEC | ✓ |
| WJEC LINEAR | ✓ |
| CCEA | ✓ |

> **KEY POINT**
>
> A **factor** is a number that divides exactly into another number. It is a divisor. A **multiple** is a number that can be divided exactly by another number.

Some numbers have **common factors** and **common multiples**. For example:

- 8 is a factor of 24 (24 ÷ 8 = 3) and 40 (40 ÷ 8 = 5) so it is a common factor of both.
- 10 is a multiple of 2 (2 × 5 = 10), 5 (5 × 2 = 10), and 10 (10 × 1 = 10) so it is a **common multiple** of all of them.

> **KEY POINT**
>
> Highest Common Factor (**HCF**) is the highest factor common to two or more numbers.
>
> Least Common Multiple (**LCM**) is the lowest multiple common to two or more numbers.

For example:

- 30 has factors 1, 2, 3, 5, 6, 10, 15 and 30
  45 has factors 1, 3, 5, 9, 15 and 45
  ∴ 3, 5 and 15 are common factors of 30 and 45, but their HCF = 15

- 7 has multiples 7, 14, 21, 28, 35, ...
  4 has multiples 4, 8, 12, 16, 20, 24, 28, 32, ...
  ∴ 28 is the lowest multiple common to 4 and 7, so their LCM = 28

> The times table of a number is a list of its multiples.

# Prime numbers

| | |
|---|---|
| AQA A | ✓ |
| AQA B | ✓ |
| EDEXCEL A | ✓ |
| EDEXCEL B | ✓ |
| OCR A | ✓ |
| OCR B | ✓ |
| WJEC | ✓ |
| WJEC LINEAR | ✓ |
| CCEA | ✓ |

> **KEY POINT**
>
> A **prime number** has only two factors, itself and 1. The only even prime number is 2. All other prime numbers are odd.

For example:

- 17 has factors 1 and 17 so it is a prime number.
- 15 has factors 1, 3, 5 and 15 so it is not a prime number.

> **KEY POINT**
>
> A factor that is a prime number is a **prime factor**. All positive integers can be shown to be the product of prime factors.

> The factors of every integer include itself and 1.

There are two methods of writing this down. It does not matter which you choose.

**Example**

Write 18 and 45 as products of their prime factors.
Then find the HCF and LCM of 18 and 45.

Method 1                          Method 2

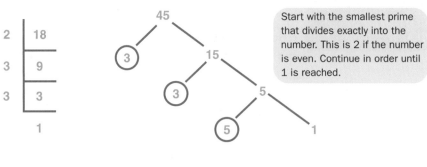

Start with the smallest prime that divides exactly into the number. This is 2 if the number is even. Continue in order until 1 is reached.

$18 = 2 \times 3 \times 3$                   $45 = 3 \times 3 \times 5$

18 and 45 have factors $3 \times 3$ in common.

∴ HCF = $3 \times 3 = 9$

To find the LCM, you only need to count the common factors once, but all other factors must be included.

∴ LCM = $2 \times 3 \times 3 \times 5 = 90$

**PROGRESS CHECK**

**Without using a calculator**

1. Work out:
    **(a)** $5(8 - 3) - 2(6 \div 2)$
    **(b)** $4(1 + 5) \times 3 + 7$
2. Work out:
    **(a)** $6 + (-2) - (-3)$
    **(b)** $-15 \times -2$
    **(c)** $64 \div -16$
    **(d)** $-144 \div 6$
3. Write out the prime factors and find the HCF and LCM of:
    **(a)** 75 and 120
    **(b)** 15, 60 and 108

3. (a) $75 = 3 \times 5 \times 5$, $120 = 2 \times 2 \times 2 \times 3 \times 5$, HCF = 15, LCM = 600
   (b) $15 = 3 \times 5$, $60 = 2 \times 2 \times 3 \times 5$, $108 = 2 \times 2 \times 3 \times 3 \times 3$, HCF = 3, LCM = 540
2. (a) 7
   (b) 30
   (c) -4
   (d) -24
1. (a) 19
   (b) 79

# 1.2 Powers and roots

**LEARNING SUMMARY**

After studying this section, you should be able to understand:

- index notation
- square roots and cube roots
- index laws
- standard index form

## Index notation

| AQA UNITISED | ✓ |
| AQA LINEAR | ✓ |
| EDEXCEL A | ✓ |
| EDEXCEL B | ✓ |
| OCR A | ✓ |
| OCR B | ✓ |
| WJEC UNITISED | ✓ |
| WJEC LINEAR | ✓ |
| CCEA | ✓ |

**KEY POINT**

Using **index** or **power** notation is a shorthand way of writing numbers.

A positive index means that you multiply the integer by itself that number of times.

For example:

- $4^2 = 4 \times 4$ ← 2 is the index

  This is 4 squared or 4 to the power of 2.

- $4^3 = 4 \times 4 \times 4$ ← 3 is the index

  This is 4 cubed or 4 to the power of 3.

> $4^3$ does not mean $4 \times 3$. Work out the value and you will see why!

> Use the $\frac{1}{x}$ or $x^{-1}$ calculator key to find the reciprocal of a number.

A negative index means that you use the **reciprocal** (divide the number into 1) and the index becomes positive.

For example:

- $7^{-3} = \dfrac{1}{7^3}$

- $5^{-4} = \dfrac{1}{5^4}$

- $\dfrac{1}{3^2} = 3^{-2}$

If the index is zero, the result is always 1. For example, $8^0 = 1$, $126^0 = 1$, $x^0 = 1$

A fraction index means a root. For example, $9^{\frac{1}{2}} = \sqrt{9}$, $27^{\frac{1}{3}} = \sqrt[3]{27}$, $p^{\frac{2}{5}} = \sqrt[5]{p^2}$

## Square roots and cube roots

| AQA UNITISED | ✓ |
| AQA LINEAR | ✓ |
| EDEXCEL A | ✓ |
| EDEXCEL B | ✓ |
| OCR A | ✓ |
| OCR B | ✓ |
| WJEC UNITISED | ✓ |
| WJEC LINEAR | ✓ |
| CCEA | ✓ |

### Square root ($\sqrt{\ }$)

**KEY POINT**

Taking the **square root** of a number is the opposite of squaring.

For example:

The square root of 25 ($\sqrt{25}$) is both +5 and -5 as $(+5)^2$ and $(-5)^2$ give 25.

Write this as $\sqrt{25} = \pm 5$

## Cube root ($\sqrt[3]{\phantom{x}}$)

> **KEY POINT**
>
> Taking the **cube root** of a number is the opposite of cubing.

For example:

The cube root of 27 ($\sqrt[3]{27}$) is +3 as $(+3)^3$ gives 27. Write this as $\sqrt[3]{27} = 3$.

The cube root of 27 is not -3 as $(-3)^3$ gives -27.

> It helps if you can remember the squares of numbers up to 15 and the cubes of numbers up to 10, together with their square and cube roots.

**Examples**

Find:

**(a)** $\sqrt{36}$

$\sqrt{(6 \times 6)}$ and $\sqrt{(-6 \times -6)}$ so $\sqrt{36} = \pm 6$

**(b)** $\sqrt[3]{125}$

$\sqrt[3]{(5 \times 5 \times 5)} = 5$

# Index laws

| | |
|---|---|
| AQA UNITISED | ✓ |
| AQA LINEAR | ✓ |
| EDEXCEL A | ✓ |
| EDEXCEL B | ✓ |
| OCR A | ✓ |
| OCR B | ✓ |
| WJEC UNITISED | ✓ |
| WJEC LINEAR | ✓ |
| CCEA | ✓ |

You need to learn the following **index laws** or rules. The base integer has to be the same when index laws are applied.

- **Add powers when multiplying**

  $3^2 \times 3^5 = 3^{(2+5)} = 3^7$

  $x^3 \times x^2 \times x^4 = x^{(3+2+4)} = x^9$

- **Subtract powers when dividing**

  $6^3 \div 6 = 6^{(3-1)} = 6^2$     If no power is given, it is assumed the index is 1 so $6 = 6^1$

  $y^{10} \div y^3 = y^{(10-3)} = y^7$

- **Multiply powers when raising one power to another power**

  $(4^2)^3 = 4^{(2 \times 3)} = 4^6$

  $(n^4)^5 = n^{(4 \times 5)} = n^{20}$

> If a question asks for a value, do not leave it in index form.

> Work with numbers and letters separately if they are in the same expression.
> For example
> $2a^3 \times 5a^2 = 10a^5$

**Examples**

Work out:

**(a)** $5^2 \times 5$

$= 5^{(2+1)} = 5^3 = 5 \times 5 \times 5 = 125$

**(b)** $10^5 \div 10^4$

$= 10^{(5-4)} = 10^1 = 10$

**(c)** $(3^2)^3$

$= 3^{(2 \times 3)} = 3^6 = 3 \times 3 \times 3 \times 3 \times 3 \times 3 = 729$

# Standard index form

AQA UNITISED ✓
AQA LINEAR ✓
EDEXCEL A ✓
EDEXCEL B ✓
OCR A ✓
OCR B ✓
WJEC UNITISED ✓
WJEC LINEAR ✓
CCEA ✓

For standard index form on a calculator, see page 41.

**KEY POINT**

**Standard index form** is a shorthand way of writing very large or very small numbers. This is done by converting them into the form $a \times 10^n$ where $a$ is any number between 1 and 10 and $n$ is a power of 10.

**Examples**

1. Write in standard index form:
   (a) $2\,600\,000 = 26 \times 100\,000$
   $= 2.6 \times 10^6$

   (b) $0.000\,000\,0735 = 735 \div 10^{10}$
   $= 735 \times 10^{-10}$
   $= 7.35 \times 10^{-8}$

2. Write the following as ordinary numbers:
   (a) $1.72 \times 10^7 = 1.72 \times 10 \times 10 \times 10 \times 10 \times 10 \times 10 \times 10$
   $= 17\,200\,000$

   (b) $3.21 \times 10^{-8} = 3.21 \div 10^8$
   $= 0.000\,000\,0321$

3. Write the answer in standard index form:
   $$\frac{7 \times 10^3 \times 10^{-1} \times 6.1 \times 10^6}{2 \times 10^4}$$
   $$= \frac{(7 \times 6.1) \times (10^3 \times 10^{-1} \times 10^6)}{2 \times 10^4}$$
   $$= \frac{42.7 \times 10^8}{2 \times 10^4}$$
   $$= 21.35 \times 10^4$$
   $$= 2.135 \times 10^5$$

Collect together numbers and powers of 10 and work them out separately.

**PROGRESS CHECK**

**Without using a calculator**

1. Simplify and leave your answer in index form:
   (a) $(4^{-3})^2$    (b) $12^7 \div 12^4$
   (c) $2a^6 \times 3a^2 \times a$    (d) $(n^8)^{\frac{1}{4}}$

2. Find the value of:
   (a) $6.8^0$    (b) $3^2 \times 3^4$
   (c) $5^{-2}$    (d) $(2^6)^{\frac{1}{2}}$

3. Give your answer in standard index form:
   (a) $0.00275$    (b) $4\,215\,000$
   (c) $\dfrac{1}{\sqrt{25}}$    (d) $(2.76 \times 10^{-3}) \div (3 \times 10^{-5})$

3. (a) $2.75 \times 10^{-3}$ (b) $4.215 \times 10^6$ (c) $\frac{1}{5} = 0.2 = 2 \times 10^{-1}$ (d) $9.2 \times 10$
2. (a) 1 (b) 729 (c) 0.04 = $\frac{1}{25}$ (d) 8
1. (a) $4^{-6}$ (b) $12^3$ (c) $6a^9$ (d) $n^2$

# 1.3 Fractions

After studying this section, you should be able to understand:

- equivalent fractions
- improper fractions and mixed numbers
- working with fractions

## Equivalent fractions

**KEY POINT**

**Equivalent fractions** have equal value, but different form.

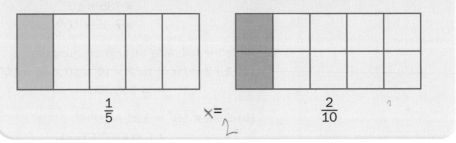

$$\frac{1}{5} \qquad \times = \qquad \frac{2}{10}$$

To find equivalent fractions, multiply or divide the top number (**numerator**) and the bottom number (**denominator**) by the same factor.

For example:

$$\frac{3 \times 2}{5 \times 2} = \frac{6}{10}$$

$$\frac{20 \div 5}{25 \div 5} = \frac{4}{5}$$

**KEY POINT**

Dividing by a common factor is called cancelling to lowest terms.

It is quicker to cancel by the highest common factor, but you should reach the same answer if you cancel in stages.

For example:

$$\frac{60 \div 4}{128 \div 4} = \frac{15}{32} \qquad \leftarrow \boxed{\text{HCF} = 4}$$

but

$$\frac{60 \div 2}{128 \div 2} = \frac{30 \div 2}{64 \div 2} = \frac{15}{32} \qquad \leftarrow \boxed{\begin{array}{l}\text{Cancel by 2 first and then}\\\text{cancel by 2 again}\end{array}}$$

**Examples**

**1.** Simplify $\dfrac{360}{480}$ into its lowest terms:

$$\dfrac{360 \div 10}{480 \div 10} = \dfrac{36 \div 12}{48 \div 12} = \dfrac{3}{4}$$

> The HCF of 360 and 480 = 120 but it is easier to cancel by 10 first

**2.** Use equivalent fractions to put these fractions in order, starting with the lowest:

$$\dfrac{4}{5}, \dfrac{1}{60}, \dfrac{7}{12}, \dfrac{6}{15}, \dfrac{2}{3}$$

The lowest common denominator of the fractions = 60. Multiply the denominators by a factor to give 60 in each case.

> Remember that the numerator and denominator must be multiplied by the same factor.

$$\dfrac{4 \times 12}{5 \times 12} = \dfrac{48}{60}$$

$$\dfrac{1 \times 1}{60 \times 1} = \dfrac{1}{60}$$

$$\dfrac{7 \times 5}{12 \times 5} = \dfrac{35}{60}$$

$$\dfrac{6 \times 4}{15 \times 4} = \dfrac{24}{60}$$

$$\dfrac{2 \times 20}{3 \times 20} = \dfrac{40}{60}$$

In order from the lowest: $\dfrac{1}{60}, \dfrac{24}{60}, \dfrac{35}{60}, \dfrac{40}{60}, \dfrac{48}{60}$

Always give the answer in the original terms: $\dfrac{1}{60}, \dfrac{6}{15}, \dfrac{7}{12}, \dfrac{2}{3}, \dfrac{4}{5}$

## Improper fractions and mixed numbers

| | |
|---|---|
| AQA UNITISED | ✓ |
| AQA LINEAR | ✓ |
| EDEXCEL A | ✓ |
| EDEXCEL B | ✓ |
| OCR A | ✓ |
| OCR B | ✓ |
| WJEC UNITISED | ✓ |
| WJEC LINEAR | ✓ |
| CCEA | ✓ |

**KEY POINT**

An **improper fraction** has a numerator larger than the denominator.
For example: $\dfrac{11}{3}, \dfrac{15}{4}, \dfrac{17}{2}$

A **mixed number** (or **mixed fraction**) consists of a whole number with a fraction.
For example: $1\frac{2}{7}, 4\frac{1}{5}, 6\frac{2}{3}$

Improper fractions can be converted into mixed numbers by dividing the numerator by the denominator.

For example:

$$\dfrac{11}{3} = 3 \text{ (remainder 2)} = 3\tfrac{2}{3}$$

Mixed numbers can be converted into improper fractions. Multiply the whole number by the denominator, add it to the numerator and put the total over the denominator.

For example:

$$1\tfrac{2}{7} = \dfrac{(1 \times 7) + 2}{7} = \dfrac{9}{7}$$

# Working with fractions

AQA UNITISED ✓
AQA LINEAR ✓
EDEXCEL A ✓
EDEXCEL B ✓
OCR A ✓
OCR B ✓
WJEC UNITISED ✓
WJEC LINEAR ✓
CCEA ✓

## Adding and subtracting fractions

**KEY POINT**

To add or subtract fractions, first find the lowest common denominator (their LCM) and convert to equivalent fractions.

- Adding fractions:
$$\frac{2}{3} + \frac{1}{6}$$

> LCM = 6 so multiply the numerator and denominator in $\frac{2}{3}$ by 2 to get the equivalent fraction

$$= \frac{4}{6} + \frac{1}{6} = \frac{5}{6}$$

> You must not add numerators and denominators.
> $\frac{2}{3} + \frac{1}{6} \neq \frac{3}{9}$

- Subtracting fractions:
$$\frac{4}{5} - \frac{3}{10}$$

> LCM = 10 so multiply the numerator and denominator in $\frac{4}{5}$ by 2 to get the equivalent fraction

$$= \frac{8}{10} - \frac{3}{10} = \frac{5}{10} = \frac{1}{2}$$

> Give answer in lowest terms, so cancel by 5

## Adding and subtracting mixed numbers

**Examples**

Find:

(a) $2\frac{4}{7} + 1\frac{1}{5}$

$$= 3 + \left(\frac{4}{7} \begin{smallmatrix} \times 5 \\ \times 5 \end{smallmatrix} + \frac{1}{5} \begin{smallmatrix} \times 7 \\ \times 7 \end{smallmatrix}\right) = 3 + \left(\frac{20}{35} + \frac{7}{35}\right) = 3\frac{27}{35}$$

> LCM = 35

> To avoid making mistakes when subtracting, change mixed numbers to improper fractions first.

(b) $4\frac{2}{3} - 2\frac{3}{4}$

$$= \frac{14}{3} - \frac{11}{4}$$

$$= \frac{14}{3} \begin{smallmatrix} \times 4 \\ \times 4 \end{smallmatrix} - \frac{11}{4} \begin{smallmatrix} \times 3 \\ \times 3 \end{smallmatrix} = \frac{56}{12} - \frac{33}{12}$$

> LCM = 12

$$= \frac{23}{12}$$

$$= 1\frac{11}{12}$$

> You can do the calculation without changing to improper fractions first but make sure you work through the calculation carefully.

(c) $3\frac{7}{10} - 2\frac{1}{4}$

$$= 1 + \left(\frac{7}{10} \begin{smallmatrix} \times 2 \\ \times 2 \end{smallmatrix} - \frac{1}{4} \begin{smallmatrix} \times 5 \\ \times 5 \end{smallmatrix}\right) = 1 + \left(\frac{14}{20} - \frac{5}{20}\right) = 1\frac{9}{20}$$

> LCM = 20

## Multiplying fractions

**KEY POINT**

To multiply two fractions, multiply the numerators and multiply the denominators. You may cancel first if there are common factors, otherwise cancel at the end.

**Examples**

Find:

**(a)** $\dfrac{3}{5} \times \dfrac{2}{7}$

$= \dfrac{3 \times 2}{5 \times 7}$

$= \dfrac{6}{35}$

**(b)** $\dfrac{6}{7} \times \dfrac{14}{25}$

$= \dfrac{6 \times \cancel{14}^{2}}{\cancel{7}_{1} \times 25}$ ← Cancel by 7

$= \dfrac{6 \times 2}{1 \times 25}$

$= \dfrac{12}{25}$

## Dividing fractions

Division is the opposite or inverse of multiplication.

For example, if you divide by 5 the result is the same as multiplying by $\dfrac{1}{5}$

$25 \div 5 = 5 \qquad 25 \times \dfrac{1}{5} = \dfrac{25}{5} = 5$

**KEY POINT**

Dividing by a fraction gives the same result as multiplying by its reciprocal.

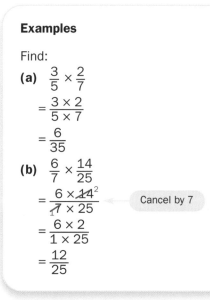
All integers can be written over 1 to form a fraction.

**Examples**

Find:

**(a)** $\dfrac{2}{7} \div \dfrac{1}{3}$

$= \dfrac{2}{7} \times \dfrac{3}{1}$ ← Turn divisor upside down and multiply

$= \dfrac{6}{7}$

**(b)** $1\dfrac{4}{5} \div 2\dfrac{1}{3}$

$= \dfrac{9}{5} \div \dfrac{7}{3}$

$= \dfrac{9}{5} \times \dfrac{3}{7}$

$= \dfrac{27}{35}$

Always change mixed numbers to improper fractions before multiplying or dividing.

**PROGRESS CHECK**

**1** Write these fractions in their lowest terms:

(a) $\dfrac{24}{48}$  (b) $\dfrac{18}{27}$  (c) $\dfrac{44}{77}$  (d) $\dfrac{42}{78}$

**2** Work out:

(a) $\dfrac{9}{25} + \dfrac{4}{5}$  (b) $\dfrac{5}{6} - \dfrac{1}{4}$  (c) $2\dfrac{3}{4} + 1\dfrac{2}{3}$  (d) $3\dfrac{1}{7} - \dfrac{5}{14}$

**3** Work out:

(a) $\dfrac{4}{15} \times 3$  (b) $\dfrac{2}{7} \div \dfrac{2}{3}$  (c) $1\dfrac{2}{5} \times 3\dfrac{1}{4}$  (d) $2\dfrac{1}{4} \div \dfrac{5}{6}$

1. (a) $\frac{1}{2}$ (b) $\frac{2}{3}$ (c) $\frac{4}{7}$ (d) $\frac{7}{13}$ 2. (a) $\frac{29}{25}$ or $1\frac{4}{25}$ (b) $\frac{7}{12}$ (c) $4\frac{5}{12}$ (d) $2\frac{11}{14}$ 3. (a) $\frac{4}{5}$ (b) $\frac{3}{7}$ (c) $4\frac{11}{20}$ (d) $2\frac{7}{10}$

# 1.4 Decimals

**After studying this section, you should be able to understand:**

- place value
- working with decimals
- converting fractions to decimals
- converting decimals to fractions

## Place value

| AQA UNITISED | ✓ |
|---|---|
| AQA LINEAR | ✓ |
| EDEXCEL A | ✓ |
| EDEXCEL B | ✓ |
| OCR A | ✓ |
| OCR B | ✓ |
| WJEC UNITISED | ✓ |
| WJEC LINEAR | ✓ |
| CCEA | ✓ |

**Decimals** are numbers based on 10. They consist of an integer followed by a decimal point and a decimal fraction.

Integer → **102.35** ← Decimal fraction

Decimal point

**KEY POINT**

The **place value** of each digit depends on its position in reference to the decimal point. Decimal place is usually shortened to d.p.

| 1000s | 100s | 10s | 1s | | $\frac{1}{10s}$ | $\frac{1}{100s}$ | $\frac{1}{1000s}$ |
|---|---|---|---|---|---|---|---|
| 1 | 3 | 2 | 5 | . | 2 | 3 | 4 |

As the places move to the right from the decimal point, they are divided by 10.
As the places move to the left from the decimal point, they are multiplied by 10.

**Examples**

1. Give the value of the underlined digits:
   **(a)** 10.<u>5</u>21
   5 is in the $\frac{1}{10}$ place and equals $\frac{5}{10}$
   **(b)** 3.04<u>9</u>
   9 is in the $\frac{1}{1000}$ place and equals $\frac{9}{1000}$
   **(c)** 0.000<u>7</u>
   7 is in the $\frac{1}{10000}$ place and equals $\frac{7}{10000}$

2. Put these decimals in order of size starting with the smallest:
   2.273   2.05   2.275   2.068   2.108

   The integers before the decimal point are the same.
   Two of the decimals (2.05, 2.068) have 0 in the $\frac{1}{10}$ place so these are smaller than the other three decimals.

   2.05 is smaller than 2.068 because of the digits in the $\frac{1}{100}$ place.

   2.108 is the next decimal as it has 1 in the $\frac{1}{10}$ place.

   2.273 is smaller than 2.275 because of the digits in the $\frac{1}{1000}$ place.

   The correct order is: 2.05   2.068   2.108   2.273   2.275

Compare each place in each decimal in turn.

# Working with decimals

AQA UNITISED ✓
AQA LINEAR ✓
EDEXCEL A ✓
EDEXCEL B ✓
OCR A ✓
OCR B ✓
WJEC UNITISED ✓
WJEC LINEAR ✓
CCEA ✓

## Adding and subtracting decimals

**KEY POINT**

When adding or subtracting decimals, line up the decimal points. This makes sure the place values are correct.

**Examples**

Work out:

**(a)**  $6.36 + 42.73$

$$\begin{array}{r} 6.36 \\ +\ 42.73 \\ \hline 49.09 \\ \hline \scriptstyle 1 \end{array}$$

**(b)**  $37.83 - 5.21$

$$\begin{array}{r} 37.83 \\ -\ \ 5.21 \\ \hline 32.62 \end{array}$$

## Multiplying decimals

**KEY POINT**

When multiplying decimals, ignore the decimal points. Count the total decimal places in the numbers being multiplied and give the answer to the same total of decimal places.

**Examples**

Work out:

**(a)**  $1.35 \times 0.2$  ←  Total d.p. = 3

   ←  Multiply ignoring the decimal points

$135 \times 2 = 270$

$\therefore 1.35 \times 0.2 = 0.27$  ←  Move the digits three places to the right as there are 3 d.p. in the question. You do not need to show the final zero.

**(b)**  $0.3 \times 0.2$  ←  Total d.p. = 2

$3 \times 2 = 6$

$\therefore 0.3 \times 0.2 = 0.06$  ←  Move the digits two places to the right as there are 2 d.p. in the question. The zero in the $\frac{1}{10}$ place must be shown so that 6 is in the $\frac{1}{100}$ place.

**(c)**  $23.6 \times 1.3$

$$\begin{array}{r} 236 \\ \times \quad 13 \\ \hline 708 \\ \scriptstyle 1\ 1 \\ 2360 \\ \hline 3068 \\ \hline \scriptstyle 1 \end{array}$$

$\therefore 23.6 \times 1.3 = 30.68$

## Dividing decimals

> **KEY POINT**
>
> When dividing decimals, multiply the divisor (the number doing the dividing) by a power of ten to make it a whole number. Multiply the dividend (the number being divided) by the same power of ten. There is no need to alter the d.p. in the answer.

**Examples**

Work out:

**(a)** $0.98 \div 0.7$

$= 9.8 \div 7$     Multiply 0.7 by 10 to make it a whole number, then multiply 0.98 by 10

$= 1.4$

**(b)** $1.368 \div 0.04$

$= 136.8 \div 4$     Multiply 0.04 by 100 to make it a whole number, then multiply 1.368 by 100

$= 34.2$

**(c)** $23.04 \div 3$     It is only necessary to multiply the divisor if it is a decimal

$= 7.68$

# Converting fractions to decimals

| | |
|---|---|
| AQA UNITISED | ✓ |
| AQA LINEAR | ✓ |
| EDEXCEL A | ✓ |
| EDEXCEL B | ✓ |
| OCR A | ✓ |
| OCR B | ✓ |
| WJEC UNITISED | ✓ |
| WJEC LINEAR | ✓ |
| CCEA | ✓ |

> **KEY POINT**
>
> To convert a fraction to a decimal divide the numerator by the denominator.

It is useful to remember the decimal conversions of some common fractions such as:

$$\frac{1}{2} = 0.5 \qquad \frac{1}{4} = 0.25 \qquad \frac{1}{5} = 0.2 \qquad \frac{1}{10} = 0.1 \qquad \frac{1}{8} = 0.125$$

- A decimal that ends is called a **terminating decimal**. For example:

$$\frac{2}{5} = 2 \div 5 = 0.4$$

- A decimal that does not end, but where some numbers repeat, is called a **recurring decimal**. For example:

$$\frac{1}{3} = 1 \div 3 = 0.333333 33\ldots$$

$$\frac{3}{11} = 3 \div 11 = 0.272727 2727\ldots$$

$$\frac{2}{7} = 2 \div 7 = 0.285714285714285714\ldots$$

A shorthand way of writing a recurring decimal is to use dots to indicate the repeating pattern. For example:

$0.333333 33\ldots = 0.\dot{3}$

$0.272727 2727\ldots = 0.\dot{2}\dot{7}$

$0.285714285714285714\ldots = 0.\dot{2}8571\dot{4}$

# Converting decimals to fractions

AQA UNITISED ✓
AQA LINEAR ✓
EDEXCEL A ✓
EDEXCEL B ✓
OCR A ✓
OCR B ✓
WJEC UNITISED ✓
WJEC LINEAR ✓
CCEA ✓

All terminating decimals can be written as a fraction with an integer as the numerator and denominator. Use place value to change a decimal to a fraction.

**Examples**

Write as fractions:

**(a)** $0.9 = \dfrac{9}{10}$

**(b)** $0.07 = \dfrac{7}{100}$

**(c)** $0.39 = \dfrac{3}{10} + \dfrac{9}{100}$

$= \dfrac{30}{100} + \dfrac{9}{100}$ ⟵ $\dfrac{3}{10}$ is equivalent to $\dfrac{30}{100}$

$= \dfrac{39}{100}$

**(d)** $0.763 = \dfrac{7}{10} + \dfrac{6}{100} + \dfrac{3}{1000}$

$= \dfrac{700}{1000} + \dfrac{60}{1000} + \dfrac{3}{1000}$ ⟵ Use equivalent fractions again

$= \dfrac{763}{1000}$

Recurring decimals are converted to fractions using the method described below.

**Examples**

**1.** Write $0.\dot{3}$ as a fraction.

$0.\dot{3} = 0.333\,333\,333...$  **a**

$0.\dot{3} \times 10 = 3.333\,333\,333...$  **b**

Multiply by $10^n$ where $n$ is the total repeating digits

**b** − **a** = 3 ⟵ Subtracting eliminates the recurring digits

This means that $0.\dot{3} \times 9 = 3$

∴ $0.\dot{3} = \dfrac{3}{9} = \dfrac{1}{3}$

**2.** Write $0.\dot{2}\dot{7}$ as a fraction.

$0.\dot{2}\dot{7} = 0.272\,727\,272\,727...$  **a**

$0.\dot{2}\dot{7} \times 100 = 27.272\,727\,2727...$  **b**

There are two repeating digits so multiply by 100

**b** − **a** = 27 ⟵ Subtracting eliminates the recurring digits

This means that $0.\dot{2}\dot{7} \times 99 = 27$

∴ $0.\dot{2}\dot{7} = \dfrac{27}{99} = \dfrac{3}{11}$

**3.** Write $0.\dot{2}857\,1\dot{4}$ as a fraction.

$0.\dot{2}857\,1\dot{4} = 0.285\,714\,285\,714\,285\,714...$  **a**

$0.\dot{2}857\,1\dot{4} \times 10^6 = 285\,714.285\,714\,285\,714...$  **b**

**b** − **a** = 285\,714

This means that $0.\dot{2}857\,1\dot{4} \times 999\,999 = 285\,714$

∴ $0.\dot{2}857\,1\dot{4} = \dfrac{285\,714}{999\,999} = \dfrac{2}{7}$

If you have a fraction key [$a^b/_c$] on your calculator, you may want to use it here to put the fraction in its lowest terms.

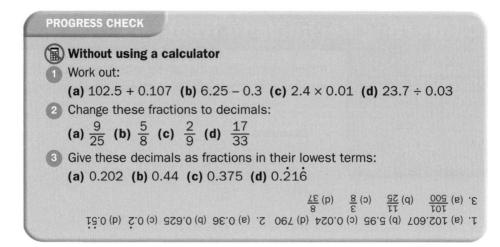

# 1.5 Percentages

**After studying this section, you should be able to understand:**

- conversion of fractions, decimals and percentages
- percentage of a quantity
- one quantity as a percentage of another
- percentage change
- reversed percentages
- simple and compound interest

**LEARNING SUMMARY**

## Conversion of fractions, decimals and percentages

AQA UNITISED ✓
AQA LINEAR ✓
EDEXCEL A ✓
EDEXCEL B ✓
OCR A ✓
OCR B ✓
WJEC UNITISED ✓
WJEC LINEAR ✓
CCEA ✓

### Percentages and fractions

A fraction with a denominator of 100 is a **percentage** (%).

For example:

- $\frac{11}{100} = 11\%$ ← This means 11 parts per 100

- $\frac{27}{100} = 27\%$

**KEY POINT**

To change a percentage to a fraction, write the percentage over 100.
Always give the fraction in its lowest terms.

**Examples**

Change to fractions:

**(a)** $9\% = \frac{9}{100}$

**(b)** $15\% = \frac{15}{100} = \frac{3}{20}$ ← Cancel by 5 to lowest terms

It is useful to remember the following percentages and fractions:

$1\% = \dfrac{1}{100}$ $\qquad$ $5\% = \dfrac{1}{20}$ $\qquad$ $10\% = \dfrac{1}{10}$ $\qquad$ $20\% = \dfrac{1}{5}$

$25\% = \dfrac{1}{4}$ $\qquad$ $33\frac{1}{3}\% = \dfrac{1}{3}$ $\qquad$ $50\% = \dfrac{1}{2}$ $\qquad$ $75\% = \dfrac{3}{4}$

**KEY POINT**

To change a fraction to a percentage, multiply by 100%.

**Examples**

Change to percentages:

**(a)** $\dfrac{3}{4} = \dfrac{3}{4} \times 100\% = \dfrac{300}{4} = 75\%$

**(b)** $\dfrac{2}{3} = \dfrac{2}{3} \times 100\% = \dfrac{200}{3} = 66\frac{2}{3}\%$

## Percentages and decimals

**KEY POINT**

To change a percentage to a decimal, divide by 100.

> Dividing by a power of 10 appears to move the decimal point to the left by the number of zeros.

**Examples**

Change to decimals:

**(a)** $63\% = \dfrac{63}{100} = 0.63$

**(b)** $35\% = \dfrac{35}{100} = 0.35$

**(c)** $8\% = \dfrac{8}{100} = 0.08$

**KEY POINT**

To change a decimal to a percentage, multiply by 100%.

> Multiplying by a power of 10 appears to move the decimal point to the right by the number of zeros.

**Examples**

Change to percentages:

**(a)** $0.43 = 0.43 \times 100\% = 43\%$

**(b)** $0.7 = 0.7 \times 100\% = 70\%$

**(c)** $0.006 = 0.006 \times 100\% = 0.6\%$

# Percentage of a quantity

| | |
|---|---|
| AQA UNITISED | ✓ |
| AQA LINEAR | ✓ |
| EDEXCEL A | ✓ |
| EDEXCEL B | ✓ |
| OCR A | ✓ |
| OCR B | ✓ |
| WJEC UNITISED | ✓ |
| WJEC LINEAR | ✓ |
| CCEA | ✓ |

**KEY POINT**

To find a percentage of a quantity, write the percentage over 100 and multiply by the quantity.

VAT stands for Value Added Tax. The percentage charged changes from time to time and country to country.

**Examples**

1. Find 20% of £200.

$$\frac{20}{{}_1\cancel{100}} \times £\cancel{200}^2 = 20 \times £2 \quad \longleftarrow \quad \boxed{\text{Cancel by 100}}$$
$$= £40$$

2. Find the VAT charged at 17.5% on a washing machine costing £398.

$$\text{VAT} = \frac{17.5}{100} \times £398 = \frac{£6965}{100}$$
$$= £69.65$$

# One quantity as a percentage of another

| | |
|---|---|
| AQA UNITISED | ✓ |
| AQA LINEAR | ✓ |
| EDEXCEL A | ✓ |
| EDEXCEL B | ✓ |
| OCR A | ✓ |
| OCR B | ✓ |
| WJEC UNITISED | ✓ |
| WJEC LINEAR | ✓ |
| CCEA | ✓ |

**KEY POINT**

To write one quantity as a percentage of another, divide the first quantity by the second quantity and multiply by 100%.

Always make sure both quantities are in the same units.

**Examples**

1. Find 45 as percentage of 180.

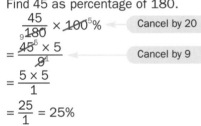

$$\frac{45}{{}_9\cancel{180}} \times \cancel{100}^5\% \quad \longleftarrow \quad \boxed{\text{Cancel by 20}}$$
$$= \frac{\cancel{45}^5 \times 5}{\cancel{9}^1} \quad \longleftarrow \quad \boxed{\text{Cancel by 9}}$$
$$= \frac{5 \times 5}{1}$$
$$= \frac{25}{1} = 25\%$$

2. Express 150m as a percentage of 120km.

$$\frac{150m}{120km} \times 100\%$$
$$= \frac{150}{120\,000} \times 100 \quad \longleftarrow \quad \boxed{1000m = 1km}$$
$$= \frac{15}{120} = 0.125\% \text{ (or } \frac{1}{8}\% \text{)} \quad \longleftarrow \quad \boxed{\text{Cancel by } 10^3}$$

3. Express 8 hours as a percentage of a day.

$$\frac{8hrs}{1 \text{ day}} \times 100\%$$
$$= \frac{8}{24} \times 100 \quad \longleftarrow \quad \boxed{\text{24hrs in a day}}$$
$$= \frac{1}{3} \times 100$$
$$= \frac{100}{3} = 33\tfrac{1}{3}\%$$

# Percentage change

Percentages are used in many real-life situations. Any of the following words for percentage increase and percentage decrease may be used:

- Percentage increase – gain, profit, inflation, surcharge, appreciation
- Percentage decrease – reduction, discount, loss, depreciation

To find the percentage change, multiply the original amount by a multiplier.

> You may be asked to find the percentage increase or decrease. Sometimes you will be asked for a final amount. Always read the question carefully.

**KEY POINT**

If there is a percentage increase, multiply by:
(1 + the percentage as a decimal)
If there is a percentage decrease, multiply by:
(1 – the percentage as a decimal)

> Another method is to find the value of the increase (or decrease) and add (or subtract) it to find the final amount. This takes longer and errors may be introduced.

**Examples**

1. A chair costs £150 plus 17.5% VAT. What is the total price?
   Multiplier = $1 + \frac{17.5}{100} = 1 + 0.175 = 1.175$ ← + because of increase
   Total cost = £150 × 1.175 = £176.25

2. A TV costing £650 is discounted by 10%. What is the sale price?
   Multiplier = $1 - \frac{10}{100} = 1 - 0.1 = 0.9$ ← – because of decrease
   Sale price = £650 × 0.9 = £585

# Reversed percentages

Sometimes you will be given an increased (or decreased) amount and asked to find the original quantity. To find the value before a percentage change, divide the original amount by a multiplier.

**KEY POINT**

To find the value before a percentage increase, divide by:
(1 + the percentage as a decimal)
To find the value before a percentage decrease, divide by:
(1 – the percentage as a decimal)

**Examples**

1. Adam gets a 3% rise in his salary. His new salary is £450 a month. What was his salary before the rise?
   Multiplier = $1 + \frac{3}{100} = 1 + 0.03 = 1.03$ ← + because of increase
   Original salary = £450 ÷ 1.03 = £436.89

2. Kizzy's bicycle costs £125 in a sale. The discount was 15%. What was the original cost?
   Multiplier = $1 - \frac{15}{100} = 1 - 0.15 = 0.85$ ← – because of decrease
   Original cost = £125 ÷ 0.85 = £147.06

> Always check that your answer makes sense!

# Simple and compound interest

| | |
|---|---|
| AQA UNITISED | ✓ |
| AQA LINEAR | ✓ |
| EDEXCEL A | ✓ |
| EDEXCEL B | ✓ |
| OCR A | ✓ |
| OCR B | ✓ |
| WJEC UNITISED | ✓ |
| WJEC LINEAR | ✓ |
| CCEA | ✓ |

**KEY POINT**

Interest is paid on money invested as savings or borrowed as a loan.

## Simple interest

**KEY POINT**

**Simple interest** is paid each year. It is not added to the original amount when the following year's interest is calculated.

Simple interest = invested or borrowed amount × time invested or borrowed in years × rate of interest as a decimal

**Example**

Find the simple interest on £1400 for 4 years at 3% per year.

Simple interest = £1400 × 4 × 0.03 = £168

Amount    Time    Interest as decimal

## Compound interest

**KEY POINT**

**Compound interest** is calculated on the amount invested or borrowed plus the interest added each year.

**Example**

£5000 is invested for 3 years at a compound interest of 3% per year. Work out the total compound interest earned.

1st year:  Investment        = £5000
           Interest          = £5000 x 0.03 = £150      *0.03 is 3% as a decimal*
           Total after 1 year = £5150

2nd year:  Investment        = £5150
           Interest          = £5150 x 0.03 = £154.50
           Total after 2 years = £5304.50

3rd year:  Investment        = £5304.50
           Interest          = £5304.50 x 0.03 = £159.14
           Total after 3 years = £5463.64

           Total compound interest  = final total – original investment
                                    = £5463.64 – £5000
                                    = £463.64

See page 42 for calculating repeated percentage increase and decrease.

**KEY POINT**

Instead of working through repeated calculations, compound interest can be calculated by using the following equation:

Final total = original amount (1 + rate of interest as a decimal)$^n$
($n$ is the number of periods of investment time)
You can then calculate the total interest.

**Example**

£5000 is invested for 3 years at a compound interest of 3% per year.
Work out the total compound interest earned.

Final total = original amount (1 + rate of interest as a decimal)$^n$
$$= £5000(1 + 0.03)^3$$
$$= £5000 \times 1.03^3$$
$$= £5463.64$$
Total compound interest = final total – original investment
$$= £5463.64 – £5000$$
$$= £463.64$$

**PROGRESS CHECK**

1. Change to fractions:
   (a) 27%  (b) 72%  (c) 5%  (d) 18%
2. Change to decimals:
   (a) 16%  (b) 7%  (c) 33%  (d) 42.5%
3. Change to percentages:
   (a) $\frac{3}{8}$  (b) $\frac{11}{15}$  (c) 0.62  (d) 0.0125
4. Find:
   (a) 3.5% of £1200
   (b) 30% of 1km
   (c) 0.5% of 500
5. Find:
   (a) 15 as a percentage of 75  (b) 40p as a percentage of £1
   (c) 40mins as a percentage of 5hrs
6. Increase:
   (a) £36 by 30%  (b) £75 by 85%  (c) £2400 by 5%
7. Decrease:
   (a) £144 by 25%  (b) £568 by 12%  (c) £2.50 by 40%
8. Find the original amount:
   (a) What was increased by 15% to give £189.75?
   (b) What was increased by 17.5% to give £324.30?
   (c) What was decreased by 30% to give £40.60?
   (d) What was decreased by 2.5% to give £81.90?
9. (a) Find the simple interest on £750 at 4% interest per year after 6 months.
   (b) A man borrowed £3500 for 2 years at 5% compound interest.
      How much is owed after 2 years?

9. (a) £15 (b) £3858.75
7. (a) £108 (b) £499.84 (c) £1.50  8. (a) £165 (b) £276 (c) £58 (d) £84
(d) £138.75 (c) £2520
5. (a) 20% (b) 40% (c) 13$\frac{1}{3}$%  6. (a) £46.80
(c) 62% (d) 1.25%  4. (a) £42 (b) 300m (c) 2.5
1. (a) $\frac{27}{100}$ (b) $\frac{18}{25}$ (c) $\frac{1}{20}$ (d) $\frac{9}{50}$  2. (a) 0.16 (b) 0.07 (c) 0.33 (d) 0.425  3. (a) 37.5% (b) 73$\frac{1}{3}$%

# 1.6 Ratio and proportion

**After studying this section, you should be able to understand:**

- simplifying ratios
- dividing a quantity in a given ratio
- proportion

## Simplifying ratios

AQA UNITISED ✓
AQA LINEAR ✓
EDEXCEL A ✓
EDEXCEL B ✓
OCR A ✓
OCR B ✓
WJEC UNITISED ✓
WJEC LINEAR ✓
CCEA ✓

> **KEY POINT**
>
> **Ratios** are used to compare quantities. They should be treated in a similar way to fractions. A colon is used to denote a ratio, e.g. 2 : 3 or 3 : 5 : 7

For example:

A bag contains 6 red balls and 3 black balls.

Ratio of red : black = 6 : 3

A ratio can be simplified or given in its lowest terms by dividing all parts by their highest common factor.

In the example above, the HCF is 3 so divide both numbers in the ratio by 3

Ratio of red : black = $\div 3 \left( \begin{array}{c} 6 : 3 \\ 2 : 1 \end{array} \right) \div 3$

> **KEY POINT**
>
> Make sure all parts of the ratio are measured in the same units.

---

**Examples**

Simplify these ratios giving them in their lowest terms:

**(a)** 125g : 250g

    = 1 : 2           HCF = 125

**(b)** 64p : £1

    = 64 : 100         Change £ to p

    = 16 : 25          HCF = 4

**(c)** 0.75km : 300m : 1.25km

    = 750 : 300 : 1250     Change km to m

    = 15 : 6 : 25        HCF = 50

---

> **KEY POINT**
>
> Ratios may sometimes be given in the form 1 : $n$ where $n$ represents a number. This is used to write scales for maps.

**Examples**

1. Write the ratio 5 : 6 in the form 1 : $n$

$$5 : 6 = \frac{5}{5} : \frac{6}{5}$$ ◄── Divide both parts of the ratio by the first amount

$$= 1 : 1\frac{1}{5}$$

$$= 1 : 1.2$$

2. On a map 2cm represents 1km. Write this scale in the form 1 : $n$

Scale = 2cm : 1km

= 2cm : 1000m ◄── Change km to m

= 2cm : 100 000cm ◄── Change m to cm

= 1 : 50 000 ◄── Divide both sides of the ratio by 2

# Dividing a quantity in a given ratio

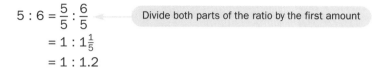

| | |
|---|---|
| AQA UNITISED | ✓ |
| AQA LINEAR | ✓ |
| EDEXCEL A | ✓ |
| EDEXCEL B | ✓ |
| OCR A | ✓ |
| OCR B | ✓ |
| WJEC UNITISED | ✓ |
| WJEC LINEAR | ✓ |
| CCEA | ✓ |

**KEY POINT**

To divide a quantity in a given ratio, first find the total number of parts.
Divide the quantity by this total, to give the value of a single share.
Multiply each part of the ratio by the value of the single share.

**Examples**

1. Divide 120 in the ratio 5 : 3

Total number of shares = 5 + 3 = 8
Value of one share = 120 ÷ 8 = 15
Value of five shares = 5 × 15 = 75
Value of three shares = 3 × 15 = 45
∴ 120 divided in the ratio 5 : 3 = 75 : 45

2. A grandmother leaves £21 000 in her will to be shared by her four grandchildren in the ratio 3 : 5 : 4 : 2. How much will each grandchild receive?

Total number of shares = 3 + 5 + 4 + 2 = 14
Value of one share = £21 000 ÷ 14 = £1500
Value of three shares = £1500 × 3 = £4500
Value of five shares = £1500 × 5 = £7500
Value of four shares = £1500 × 4 = £6000
Value of two shares = £1500 × 2 = £3000
∴ grandchildren receive £4500, £7500, £6000, £3000

> Check your working by adding up all the shares. The total should equal the original amount.

3. A water jug fills two glasses in the ratio 5 : 3
If the larger glass holds 250ml, how much does the small glass hold?
What is the total capacity of the jug?

The large glass holds 250ml and is 5 parts from the ratio.
∴ 1 part = 250 ÷ 5 = 50ml
So the small glass holds 3 × 50ml = 150ml
and the jug holds 8 parts = 8 × 50ml = 400ml

# Proportion

Convert ratio to form $n : 1$ where $n$ is the multiplier.

**KEY POINT**

Quantities are said to be in **proportion** if their ratio stays the same as they increase or decrease. An amount may increase or decrease by a given ratio.

**Examples**

1. Increase 35 in the ratio 5 : 2

$$5 : 2 = \frac{5}{2} : \frac{2}{2}$$ ← Divide both parts of the ratio by the second amount

$$= 2.5 : 1$$ ← 2.5 is the multiplier

∴ 35 increased in ratio 5 : 2 = 2.5 × 35 = 87.5

2. Decrease 24 in the ratio 4 : 5

$$4 : 5 = \frac{4}{5} : \frac{5}{5}$$ ← Divide both parts of the ratio by the second amount

$$= 0.8 : 1$$ ← 0.8 is the multiplier

∴ 24 decreased in ratio 4 : 5 = 0.8 × 24 = 19.2

## Direct proportional change

**KEY POINT**

**Direct proportional change** is when two or more quantities increase in the same ratio.

**Example**

12 pens cost 90p. How much will 16 pens cost?

Find the cost of one item, then multiply by the quantity required. This is known as the **unitary method**.
One pen costs 90p ÷ 12 = 7.5p
16 pens cost 16 × 7.5p = 120p
∴ 16 pens cost £1.20 ← It is more sensible to give cost in £ and p

As the number of pens increases, so does the cost in direct proportion.

# Inverse or indirect proportional change

**KEY POINT**

**Inverse proportional change** is when one quantity increases as the other decreases in the same ratio.

**Example**

3 men take 5 days to paint a house. How long would 4 men take to paint the same house?

1 man would take 3 times as long as 3 men to complete the same task.
So, 1 man would take 3 × 5 days = 15 days
As the number of men decreases, the time taken increases in inverse proportion.

So, 4 men would take $\frac{15}{4}$ days

∴ 4 men would take $3\frac{3}{4}$ days.

**PROGRESS CHECK**

1. Simplify these ratios:
   (a) 108 : 81    (b) 1 day : 4hrs
   (c) 36m : 48m : 1.44km    (d) 2mins: 5mins: 90secs
2. (a) Divide 24hrs in the ratio 1 : 3 : 4
   (b) Divide £108 in the ratio 2 : 3 : 5
   (c) An orchard has white cherry trees and black cherry trees planted in the ratio 3 : 5. There are 30 black cherry trees. How many white cherry trees are there?
   (d) A triangle's sides are in the ratio 5 : 12 : 13. The perimeter is 60cm. Find the length of each side of the triangle.
3. Increase these quantities in the given ratio:
   (a) 12 in the ratio 4 : 3    (b) £96 in the ratio 7 : 3
4. Decrease these quantities in the given ratio:
   (a) 40 in the ratio 2 : 5    (b) £77 in the ratio 3 : 4
5. Ben drives 290 miles in 5 hours of steady driving. How far can he drive in 7 hours at the same speed?
6. A fruit farmer has 10 helpers to pick his apples. They will take 12 days. How many helpers would he need to pick the apples in 10 days?

# 1.7 Approximations

> **LEARNING SUMMARY**
>
> After studying this section, you should be able to understand:
> - rounding numbers
> - estimating results
> - rational numbers, irrational numbers and surds

## Rounding numbers

| | |
|---|---|
| AQA UNITISED | ✓ |
| AQA LINEAR | ✓ |
| EDEXCEL A | ✓ |
| EDEXCEL B | ✓ |
| OCR A | ✓ |
| OCR B | ✓ |
| WJEC UNITISED | ✓ |
| WJEC LINEAR | ✓ |
| CCEA | ✓ |

> **KEY POINT**
>
> If an approximate value is required, a number may be rounded.

### Rounding integers to powers of 10

A number may be rounded to the nearest 10, 100, 1000, etc.

For example, 75 378 people attend a football match at Old Trafford.

75 378 rounded to the nearest 1000 becomes 75 000 ← Look at the thousands, 5378 is nearer 5000 than 6000

75 378 rounded to the nearest 100 becomes 75 400 ← Look at the hundreds, 378 is nearer 400 than 300

75 378 rounded to the nearest 10 becomes 75 380 ← Look at the tens, 78 is nearer 80 than 70

> **KEY POINT**
>
> It is usual to round up 'halfway' numbers. For example, if 25 is to be rounded to the nearest 10, round up to 30.

### Rounding to a given number of decimal places

> **KEY POINT**
>
> Decimal places are counted to the right from the decimal point. Use d.p. as an abbreviation for decimal place. For example, 3.2 has 1 d.p. and 0.002 564 has 6 d.p.

**Example**

Write 36.136 to 1 decimal place and 2 decimal places.

The digit in the first decimal place is 1. The digit in the second decimal place is 3. The digit in the third decimal place is 6.

∴ 36.136 = 36.1 to 1 d.p. ← 3 is less than 5 so do not round up

36.136 = 36.14 to 2 d.p. ← 6 is greater than 5 so round up

# Rounding to a given number of significant figures

> **KEY POINT**
>
> **Significant figures** are counted from the number's first digit from the left, disregarding zeros.
>
> For example, 3.2 has 2 significant figures. The most significant figure is 3. 0.002 564 has 4 significant figures. The most significant figure is 2.
>
> Use s.f. or sig. fig. as abbreviations for significant figures.

### Examples

Write the following numbers to the number of significant figures given in brackets.

**(a)** 30 645   (3 s.f.)

Count 3 places from the most significant figure. The next digit is 4.
As 4 is less than 5, do not round up.

∴ 30 645 = 30 600 to 3 s.f.

> Replace the 4 and 5 with zeros to keep the place value of the remaining digits

**(b)** 0.001 07   (2 s.f.)

Count 2 places from the most significant figure. The next digit is 7.
As 7 is greater than 5, round up.

∴ 0.001 07 = 0.0011 to 2 s.f.

> The zeros in the first 2 decimal places after the decimal point keep the place value of the remaining digits. Do not put a zero in the fifth decimal place.

# Degree of accuracy

> **KEY POINT**
>
> An answer should be given to the same **degree of accuracy** as the question, unless instructed otherwise.

### Example

A room measures 12.5m by 16.7m. What area of carpet is needed to cover the floor?

Area of floor = $12.5 \times 16.7 = 208.75 m^2$
The original measurements were given to 1 d.p. It is sensible to give the area to the same degree of accuracy.

∴ Area of carpet = $208.8 m^2$

# Upper and lower bounds

> **KEY POINT**
>
> When you are given an amount that has been rounded, it is useful to know the highest and lowest values it could have originally had. These are called the **upper and lower bounds**.

**Example**

The measurements of a piece of paper are given as 300mm by 210mm correct to the nearest 10mm. What are the highest and lowest possible values of these measurements?

300mm could be the result of rounding 295mm up to the nearest 10 or rounding 304.$\dot{9}$ down to the nearest 10.
210mm could be the result of rounding 205mm up to the nearest 10 or rounding 214.$\dot{9}$ down to the nearest 10.

The upper and lower bounds of the measurements are:

$295\text{mm} \leqslant 300\text{mm} < 305\text{mm}$

$205\text{mm} \leqslant 210\text{mm} < 215\text{mm}$

> It is easier to use < 305mm and < 215mm than recurring decimals as the upper bounds

> See page 61 for the symbols used to represent inequalities.

# Estimating results

| | |
|---|---|
| AQA UNITISED | ✓ |
| AQA LINEAR | ✓ |
| EDEXCEL A | ✓ |
| EDEXCEL B | ✓ |
| OCR A | ✓ |
| OCR B | ✓ |
| WJEC UNITISED | ✓ |
| WJEC LINEAR | ✓ |
| CCEA | ✓ |

Estimating a result is useful. It indicates whether your answer makes sense.

**KEY POINT**

To estimate a result, approximate or round each number in the calculation. This means you can work out an estimated answer quickly.

Use ≈ to mean 'approximately equal to'.

**Examples**

1. Estimate 519 ÷ 23

   $519 \div 23 \approx 520 \div 20 = 26$ ← Round both numbers to the nearest 10
   $\therefore 519 \div 23 \approx 26$

   The actual answer, worked out by long division, or a calculator, is 22.565 correct to 3 d.p. Both answers are in the same order of magnitude (i.e. roughly the same size).

2. Estimate $\dfrac{91.4 \times 3.8}{9.9 \times 2.35}$ then compare with the actual calculation.

   $\dfrac{91.4 \times 3.8}{9.9 \times 2.35} \approx \dfrac{\overset{9}{\cancel{90}} \times \overset{2}{\cancel{4}}}{\underset{1}{\cancel{10}} \times \underset{1}{\cancel{2}}}$   ← The most sensible approximation is to 1 s.f.

   $= \dfrac{360}{20} = 18$ or by cancelling to give $\dfrac{18}{1} = 18$

   Actual answer = 14.9 to 1 d.p.
   Both answers are in the same order of magnitude.

   > ≈ is used to show approximation. Then = can be used.

3. Jacob multiplies 413 by 0.025 and gets 0.103 25 as his answer. Use approximation to find if he is correct. Give a reason for your answer.

   $413 \times 0.025 \approx 400 \times 0.03 = 12$   ← Correct both amounts to 1 s.f.

   This is not of the same order of magnitude. It is much larger.
   Actual answer = 10.325
   It looks as if Jacob has his decimal point in the wrong place.

   > This is a very good reason to estimate your answer first!

# Rational numbers, irrational numbers and surds

> **KEY POINT**
>
> A **rational number** is a number that can be written as a fraction with the numerator and denominator both integers.

Examples of rational numbers are:

$$4.75 = \frac{19}{4} \qquad -5 = \frac{-5}{1} \qquad 2\frac{2}{3} = \frac{8}{3} \qquad \sqrt{49} = \frac{7}{1} \qquad 0.625 = \frac{5}{8} \qquad 0.272727\ldots = \frac{3}{11}$$

> **KEY POINT**
>
> An **irrational number** cannot be written as a fraction. Decimals that neither terminate nor recur are irrational numbers.

Examples of irrational numbers are:

$$\sqrt{2} \qquad 1 + \sqrt{5} \qquad \sqrt[3]{7} \qquad \pi$$

> **KEY POINT**
>
> A **surd** is a square root that does not give an exact result, so all surds are irrational numbers.

For example, the following numbers are all surds:

$$\frac{\sqrt{2}}{\sqrt{5}} \qquad 5 + 3\sqrt{7} \qquad \sqrt[3]{7}$$

The square root of a prime number is a surd and so is an irrational number.

## Simplifying surds

Surds can be manipulated so that an expression is simplified. The following rules should be followed when simplifying surds ($\sqrt{a}$ and $\sqrt{b}$ are surds):

- $\sqrt{a} \times \sqrt{b} = \sqrt{ab}$
- $\dfrac{\sqrt{a}}{\sqrt{b}} = \sqrt{\dfrac{a}{b}}$
- $c\sqrt{a} + d\sqrt{a} = (c + d)\sqrt{a}$
- $c\sqrt{a} - d\sqrt{a} = (c - d)\sqrt{a}$
- $\sqrt{a} + \sqrt{b} \neq \sqrt{a + b}$ (for example, $\sqrt{5} + \sqrt{7} = 4.88$ but $\sqrt{(5 + 7)} = \sqrt{12} = 3.46$)

**Examples**

Simplify the following. Leave the answers in surd form.

**(a)** $\sqrt{45}$

$= \sqrt{9 \times 5}$ ← Write 45 as a product of its factors

$= 3\sqrt{5}$ ← The square root of 9 is 3

**(b)** $\dfrac{2\sqrt{72}}{3}$

$= \dfrac{2\sqrt{36 \times 2}}{3}$

$= \dfrac{2 \times 6\sqrt{2}}{3}$

$= \dfrac{12\sqrt{2}}{3}$

$= 4\sqrt{2}$ ← Cancel as 3 is a common factor

**(c)** $7\sqrt{12} \div \sqrt{15}$

$= \dfrac{7\sqrt{4 \times 3}}{\sqrt{5 \times 3}}$ ← Write 12 and 15 as a product of their factors

$= \dfrac{7 \times 2\sqrt{3}}{\sqrt{5} \times \sqrt{3}}$

$= \dfrac{14\sqrt{3}}{\sqrt{5}\sqrt{3}}$

$= \dfrac{14}{\sqrt{5}}$ ← Cancel as $\sqrt{3}$ is a common factor

# Rationalising a denominator

**KEY POINT**

Rationalising a denominator means removing a surd from the denominator.

This is done by multiplying the numerator and the denominator by the surd being removed.

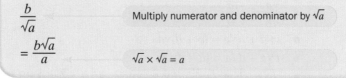

$\dfrac{b}{\sqrt{a}}$ ← Multiply numerator and denominator by $\sqrt{a}$

$= \dfrac{b\sqrt{a}}{a}$ ← $\sqrt{a} \times \sqrt{a} = a$

**Examples**

Rationalise these denominators and simplify if possible.

**(a)** $\dfrac{1}{\sqrt{5}}$ ← Multiply numerator and denominator by $\sqrt{5}$

$= \dfrac{\sqrt{5}}{5}$ ← Remember $\sqrt{5} \times \sqrt{5} = 5$

**(b)** $\dfrac{2}{\sqrt{6}}$ ← Multiply numerator and denominator by $\sqrt{6}$

$= \dfrac{2\sqrt{6}}{6}$

$= \dfrac{\sqrt{6}}{3}$ ← Cancel as 2 is a common factor

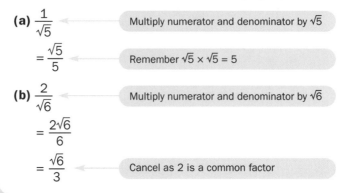

**Examples**

**(c)** $3\sqrt{\dfrac{7}{35}}$

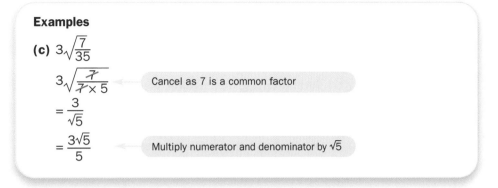

$3\sqrt{\dfrac{\cancel{7}}{\cancel{7}\times 5}}$ ← Cancel as 7 is a common factor

$=\dfrac{3}{\sqrt{5}}$

$=\dfrac{3\sqrt{5}}{5}$ ← Multiply numerator and denominator by $\sqrt{5}$

---

**PROGRESS CHECK**

**1** Round these numbers to the nearest power of 10, given in brackets:
 **(a)** 724 (100)  **(b)** 1357 (10)
 **(c)** 931 (10)  **(d)** 12 956 (1000)

**2** Correct the following to the number of d.p. or s.f. given in brackets:
 **(a)** 1.063 09 (1 d.p.)  **(b)** 0.0345 (3 d.p.)
 **(c)** 962.3812 (1 s.f.)  **(d)** 0.001 05 (2 s.f.)

**3** Find the upper and lower bounds of:
 **(a)** a weight of 120kg (rounded to nearest 10kg)
 **(b)** an audience of 1700 (rounded to the nearest 100)
 **(c)** a petrol tank of 60l (rounded to the nearest 5l)
 **(d)** a cricket crowd of 23 000 (rounded to the nearest 1000)

**4** Estimate the result without a calculator, then write down the answer using a calculator to a sensible degree of accuracy.
 **(a)** 658 × 67  **(b)** 360 ÷ 87
 **(c)** 6972 × 133  **(d)** 4792 ÷ 49

**5** Simplify the following:
 **(a)** $\sqrt{60}-\sqrt{32}$
 **(b)** $3\sqrt{24}\div\sqrt{30}$
 **(c)** $\dfrac{1+2\sqrt{3}}{\sqrt{5}}$
 **(d)** $35\sqrt{\dfrac{6}{42}}$

**1.** (a) 700 (b) 1360 (c) 930 (d) 13 000  **2.** (a) 1.1 (b) 0.035 (c) 1000 (d) 0.0011
**3.** (a) $115 \leqslant 120 < 125$ (b) $1650 \leqslant 1700 < 1750$ (c) $57.5 \leqslant 60 < 62.5$
(d) $22\,500 \leqslant 23\,000 < 23\,500$  **4.** (a) 46 200; 44 086 (b) 4; 4.14
(c) 700 000; 927 300 (d) 100; 97.80  **5.** (a) $2(\sqrt{15}-\sqrt{8})$ or $2\sqrt{15}-4\sqrt{2}$
(b) $\dfrac{6\sqrt{5}}{5}$  (c) $\dfrac{\sqrt{5}(1+2\sqrt{3})}{5}$ or $\dfrac{2\sqrt{15}+\sqrt{5}}{5}$  (d) $5\sqrt{7}$

# 1.8 Calculator use

**After studying this section, you should be able to understand:**

- calculator keys
- standard index form on a calculator
- exponential growth and decay

## Calculator keys

| | |
|---|---|
| AQA UNITISED | ✓ |
| AQA LINEAR | ✓ |
| EDEXCEL A | ✓ |
| EDEXCEL B | ✓ |
| OCR A | ✓ |
| OCR B | ✓ |
| WJEC UNITISED | ✓ |
| WJEC LINEAR | ✓ |
| CCEA | ✓ |

Decide which Mode you wish your calculator to use.

All calculators are different. You need to spend time getting used to your own calculator.

> **KEY POINT**
>
> Read your calculator's instruction manual carefully. It will help you to use your calculator efficiently and save you time in the exam. If the key you need is printed above a button use [SHIFT] to activate it.

You need to know how to use the following keys:

- **Memory**

  [M] [M–] [M+] [RCL] [STO]

  All the buttons above are examples of calculator memory keys.

  [RCL] means recall.

  [STO] means store in the memory.

- **Brackets**

  [ ( ] [ ) ]

  Brackets are used to show the order of calculating.

- **Time conversion**

  [° ' "]

  This is useful in calculations involving time which uses base 60. If your answer is displayed as a decimal this key will convert it to hours, minutes and seconds. For example:

  Your calculator displays 3.275 hours.

  [SHIFT] [° ' "] [=] gives 3 16 30 meaning 3hrs 16mins 30secs

  You wish to convert 1hr 38mins to a decimal and multiply by 3.

   [1] [° ' "] [38] [° ' "] [×] [3] [=] [SHIFT] [° ' "] gives the result 4.9

  (N.B. You may not need to use the [SHIFT] button on your calculator.)

- **Powers**

  [$x^2$] [$x^3$] [$x^y$] [$x^{-1}$] [$10^x$] [$x^\square$] or [∧]

  All the buttons above are examples of calculator keys used for powers.

  [$x^y$] is used for powers other than 2 and 3.

  [$x^{-1}$] is used for reciprocals.

  [$10^x$] is used for powers of 10.

- **Roots**

  $\sqrt{\phantom{x}}$ $\sqrt[3]{\phantom{x}}$ $\sqrt[x]{\phantom{x}}$

  The buttons above are examples of calculator keys used for roots.

  $\sqrt[x]{\phantom{x}}$ is used for roots other than square root and cube root.

- **Exponent**

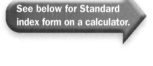

  $\boxed{\text{EXP}}$ or $\boxed{\times 10^x}$ means $\times\,10^n$ where $n$ is any power.

- **Trigonometric**

  $\boxed{\text{sin}}$ $\boxed{\text{cos}}$ $\boxed{\text{tan}}$

  Use the buttons above to solve trigonometric problems using angles.

  $\boxed{\text{sin}^{-1}}$ $\boxed{\text{cos}^{-1}}$ $\boxed{\text{tan}^{-1}}$

  These are inverse trigonometric functions found using $\boxed{\text{SHIFT}}$. Use them when finding an angle.

- **Statistical**

  $\boxed{\Sigma}$ $\boxed{x\sigma_n}$ $\boxed{\sigma_n}$ $\boxed{x\sigma_{n-1}}$ $\boxed{\bar{x}}$

  The buttons above are examples of calculator keys used in statistics. The SD Mode will be needed to use these keys. They are also used with $y$ instead of $x$.

When you are using a calculator, it is important not to round numbers too soon, as rounding is a form of estimation. Write only the final answer to an appropriate degree of accuracy.

# Standard index form on a calculator

| | |
|---|---|
| AQA UNITISED | ✓ |
| AQA LINEAR | ✓ |
| EDEXCEL A | ✓ |
| EDEXCEL B | ✓ |
| OCR A | ✓ |
| OCR B | ✓ |
| WJEC UNITISED | ✓ |
| WJEC LINEAR | ✓ |
| CCEA | ✓ |

**KEY POINT**

Use the $\boxed{\text{EXP}}$ or $\boxed{\times 10^x}$ key to input standard index form.

**Examples**

1. $1.72 \times 10^{11}$

   Enter 1.72 followed by $\boxed{\text{EXP}}$ $\boxed{11}$ $\boxed{=}$ or $\boxed{\times 10^x}$ $\boxed{11}$ $\boxed{=}$

   The answer may be given as
   $172\,000\,000\,000$ or $1.72 \times 10^{11}$ or $1.72^{11}$

   > 11 is the power of 10. This may be shown as 1.72 11, depending on your calculator

2. $3.25 \div 10^9 = 3.25 \times 10^{-9}$

   Enter 3.25 followed by $\boxed{\text{EXP}}$ $\boxed{(-)}$ $\boxed{9}$ $\boxed{=}$ or $\boxed{\times 10^x}$ $\boxed{(-)}$ $\boxed{9}$ $\boxed{=}$

   The answer may be given as
   $0.000\,000\,003\,25$ or $3.25 \times 10^{-9}$ or $3.25^{-09}$

   > -9 is the power of 10

   If your calculator has corrected the answer to fewer decimal places, the answer may be displayed as $3^{-09}$ or $3 \times 10^{-9}$

See below for Standard index form on a calculator.

See pages 93–98 Trigonometry.

# Exponential growth and decay

**Exponential growth or decay** is when a quantity grows or reduces when multiplied repeatedly by the same number, i.e. repeated percentage increase or decrease.

### KEY POINT

In order to work out a final result it is necessary to use a formula.

Repeated increase:
Final total = original amount $(1 + \text{rate of interest as a decimal})^n$

Repeated decrease:
Final total = original amount $(1 - \text{rate of interest as a decimal})^n$

($n$ is the number of periods of time. The number in the brackets is known as the multiplier.)

### Example

A house was bought 5 years ago for £325 000. The property prices have depreciated by 4% per year. What is the value of the house today to the nearest £1000?

Value of house after 5 years = £325 000 × $(1 - 0.04)^5$

Original amount     Rate     Time

$$= £325\,000 \times 0.96^5$$
$$= £264\,996.13$$
$$= £265\,000$$

### PROGRESS CHECK

**Use a calculator to answer these questions**

1. Work out the following. Give your answers correct to 2 d.p.

   (a) $\dfrac{1.85 \times 10^3}{6.7 \times 10^2}$    (b) $\dfrac{3}{(5 \div 2.1)}$

   (c) $\sqrt{(3^2 + 4^2)}$    (d) $5 + \sqrt[3]{30}$

2. A town with a population of 9500 is growing at a rate of 2% per year. Find the population after 5 years.

3. (a) A car costs £15 000 when bought new. It loses 10% of its value in each of the first 2 years. What is its value after 2 years?

   (b) Subsequently the car's value decreases at a rate of 6% per year. What is its value when it is 5 years old?

1. (a) 2.76 (b) 1.26 (c) ±5 (d) 8.11   2. 10489   3. (a) £12 150 (b) £10 091.60

# Sample GCSE questions

**1** Bella wants to buy some English breakfast tea bags.
They come in three different-sized packets: 40 tea bags costing £1.10, 80 tea bags costing £1.89, 160 tea bags costing £3.55

(a) Which is the best value? Show your working. **(1)**

(b) If Bella wants 200 tea bags, which is the best combination of packets to buy? **(2)**

(c) Are the prices of the packets in the same ratio as the number of tea bags? Give a reason for your answer. **(1)**

*It is easier to compare the cost of 1 unit in pence* →

(a) 1 teabag costs:  $110 \div 40 = 2.75p$ (40 bags)
$189 \div 80 = 2.36p$ (80 bags)
$355 \div 160 = 2.22p$ (160 bags)

∴ 160 bag size is best value.

*List all combinations to gain method marks* →

(b) 200 bags $= 5 \times 40$ costing $5 \times £1.10 = £5.50$
$= 2 \times 80 + 1 \times 40$ costing $2 \times £1.89 + £1.10 = £4.88$
$= 3 \times 40 + 1 \times 80$ costing $3 \times £1.10 + £1.89 = £5.19$
$= 1 \times 160 + 1 \times 40$ costing $£3.55 + £1.10 = £4.65$

Best combination is 1 box of 40 bags and 1 box of 160 bags.

*Divide by 40 to simplify the first ratio* →

(c) No. of teabags $= 40 : 80 : 160 = 1 : 2 : 4$
$\neq$ ratio of cost $= 110 : 189 : 355$

**2** Evaluate $\dfrac{4.23}{3.75 \times 2.31}$ using your calculator.

(a) Write down the whole display, then correct to (i) 3 s.f. (ii) 2 d.p. **(2)**

(b) Without using your calculator, show how you could check your answer. **(1)**

*Use the brackets or memory keys on your calculator* →

(a) 0.488 311 688

(i) 0.488 (3 s.f)      (ii) 0.49 (2 d.p.)

*Round each number to the nearest whole number* →

(b) $\dfrac{4.23}{3.75 \times 2.31} \approx \dfrac{4}{4 \times 2} = \dfrac{4}{8} = \dfrac{1}{2} = 0.5$

Exact answer equals estimated answer when corrected to 1 decimal place.

**3** Jacob uses his van for work. He is allowed to claim 40p tax relief for every mile he drives, up to 10 000 miles. For each mile over this limit, he can claim 25p. Last year he drove 12 500 miles.

(a) How much tax relief can he claim? **(1)**

(b) Petrol costs £1.10 per litre. Jacob's van has a petrol tank with a capacity of 12 gallons. If 1 gal = 4.5l, calculate how much it costs Jacob to fill up with petrol. **(1)**

(c) Jacob's van cost £18 455 last year. It is now valued at £11 495. What is its percentage depreciation? **(2)**

*Split the mileage into two parts* →

(a) Tax relief claimed $= (10\,000 \times 40p) + (2500 \times 25p)$
$= 400\,000p + 62\,500p$
$= £4000 + £625 = £4625$

*Show the conversion to gain the method mark* →

(b) 12 gallons $= 12 \times 4.5$ litres $= 54$ litres
∴ cost of petrol $= 54 \times £1.10 = £59.40$

*% depreciation = $\dfrac{\text{loss}}{\text{original amount}} \times 100$* →

(c) % depreciation $= \left(\dfrac{18\,455 - 11\,495}{18\,455}\right) \times 100 = \dfrac{6960}{18\,455} \times 100 = 37.7\%$

# Exam practice questions

**1** From the list of numbers below, write down... 📱
 **(a)** all multiples of 5  **(b)** all factors of 28  **(c)** all square numbers  **(d)** $\sqrt[3]{64}$  **(e)** $2^5$  **(5)**

| 4 | 7 | 25 | 14 | 15 | 32 |

**2** **(a)** Express as a product of prime factors **(i)** 36 **(ii)** 96 📱  **(4)**
 **(b)** What is the LCM and HCF of 36 and 96?  **(2)**

**3** Write each of the following as a percentage and a decimal.
 **(a)** $\frac{1}{4}$  **(b)** $\frac{3}{8}$  **(c)** $\frac{3}{4}$  **(d)** $\frac{2}{5}$  **(4)**

**4** **(a)** If next year is a leap year, what fraction of the year is February?  **(1)**
 **(b)** This year, Ben is going on a trip for 4 weeks. Express this as a fraction of a year, then convert to
  **(i)** a decimal to 3 d.p. **(ii)** standard index form  **(3)**

**5** Jasmine sees a coffee machine priced at £107.95 including VAT (VAT = 15%). When she returns to purchase it, the total price has been discounted by £20.
 **(a)** What is the sale price? What percentage decrease is this?  **(3)**
 **(b)** Calculate the cost of the coffee machine if VAT increases to 17.5% before she buys it.  **(2)**

**6** Adam wants to invest £12 500 for 2 years.
 He can either use an instant saver account, paying simple interest of 3.75% at the end of each year, or a 2 year account, paying 4.25% compound interest.
 Which account should he use and why?  **(5)**

**7** This table shows the temperatures on a December day in six cities in Europe. 📱

| London | 7°C |
|---|---|
| Milan | 12°C |
| Stockholm | -1°C |
| Reykjavik | -6°C |
| Prague | 5°C |
| Zurich | -2°C |

 **(a)** Which is the coldest place?  **(1)**
 **(b)** If you were flying between the following cities, what difference in temperature would you experience?
  **(i)** London to Prague
  **(ii)** Reykjavik to Zurich
  **(iii)** Stockholm to Milan  **(3)**

**8** **(a)** Iqra is going to France and exchanges £50 for euros. £10 will buy €12.
  How many euros does Iqra receive?  **(2)**
 **(b)** The following day, Iqra finds she has to cancel her trip and has to exchange the euros back to pounds. On the following day, €25 buys £15. How many pounds does Iqra receive?  **(2)**
 **(c)** Has Iqra lost or gained by having to exchange back from euros to pounds?  **(1)**
 **(d)** Calculate the percentage change.  **(1)**

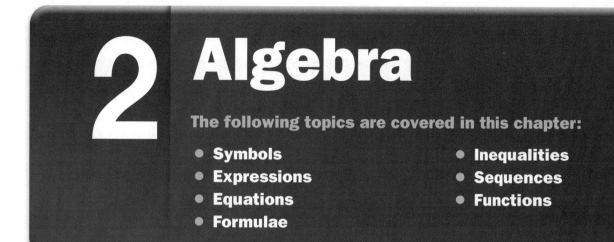

# Algebra

**The following topics are covered in this chapter:**

- Symbols
- Expressions
- Equations
- Formulae
- Inequalities
- Sequences
- Functions

# 2.1 Symbols

**LEARNING SUMMARY**

After studying this section, you should be able to understand:

- vocabulary
- using letters
- algebraic notation

## Vocabulary

| | |
|---|---|
| AQA UNITISED | ✓ |
| AQA LINEAR | ✓ |
| EDEXCEL A | ✓ |
| EDEXCEL B | ✓ |
| OCR A | ✓ |
| OCR B | ✓ |
| WJEC UNITISED | ✓ |
| WJEC LINEAR | ✓ |
| CCEA | ✓ |

**KEY POINT**

The vocabulary of algebra must be learnt. It will help you to understand what a question is asking you.

- An **expression** is an algebraic statement that has letters and numbers but no equals sign.
  For example, $2x^2 + x - 3$
- An **equation** shows two equal statements or expressions.
  For example, $6 - y = 2y + 3$
- A **sequence** is a collection of terms that follow a pattern.
  For example, 1, 3, 5, 7, …
- A **term** is part of an expression, equation or sequence.
- A **formula** is an equation used to find quantities when given certain values.
  For example, $A = l \times b$
- A **function** is a relationship between variables that depend on one another.
  For example, $f(x) = x^3 - x$
- An **identity** is an equation where what is on the left-hand side is the same as what is on the right-hand side for all possible values of $x$ ($\equiv$ is used instead of =).
  For example, $a(b + c) \equiv ab + ac$
- A **coefficient** is a constant number or letter that multiplies an algebraic term.
  For example, $3pqr$

# Using letters

**KEY POINT**

In algebra, letters are used to represent unknown numbers in expressions, formulae, equations and other algebraic functions.

Unknown **variables** in expressions and equations are often represented by $x$, $y$ and $z$. For example:

$x^2 + xy - 3y$ and $3z^2 + 3z + 1 = 0$

The + or – sign is attached to the term that follows it.
For example, in the expression $x^2 + xy - 3y$ above, $xy$ is positive (+) and $3y$ is negative (–).

Unknown numbers (**constants**) are often represented by $a$, $b$ and $c$. If these multiply a variable, they are called coefficients as in $ax^2 + by + c$.

In formulae, the unknown variables may take the initial letter of the quantities as in

$A = \pi r^2$ ← $A$ = area, $r$ = radius
$C = \pi d$ ← $C$ = circumference, $d$ = diameter

# Algebraic notation

Use the following **algebraic notation** when working with letters:

- Adding letters: $c + c + c + c$ is written as $4c$
- Multiplying letters: $a \times x$ is written as $ax$
  $3 \times x \times y$ is written as $3xy$
  $a \times a$ is written as $a^2$ ← Do not make the mistake of writing $a \times a$ as $2a$.
- A letter multiplied by zero always equals 0
- If the coefficient of a letter is 1, it is not necessary to write it down:
  $4y - 3y = 1y = y$

It is usual to write letters in alphabetical order and numbers are written in front of letters. For example, $q \times 4p \times 2s \times r$ is written as $8pqrs$

An expression or equation such as $3x + 2$ or $7x - 3 = 4$ is said to be linear because $x$ has an index (power) of 1.

An expression or equation such as $2y^2 + 5y - 4$ or $y^2 - 3y + 1 = 0$ is said to be quadratic because $y$ has an index (power) of 2.

When using powers in algebra, remember the index laws:

- $x^3 \times x^2 = x^{(3+2)} = x^5$
- $y^8 \div y^3 = y^{(8-3)} = y^5$

See page 14 for index laws.

**PROGRESS CHECK**

1. Write as one term:
   **(a)** $a \times a \times a$  **(b)** $\dfrac{b \times b \times b}{b}$  **(c)** $3 \times a \times c \times 2$  **(d)** $k \times 0$

2. **(a)** $a^2 \times a^3$  **(b)** $b^4 \div b$  **(c)** $(c^2)^3$  **(d)** $(d^4)^{\frac{1}{2}}$

1. (a) $a^3$ (b) $b^2$ (c) $6ac$ (d) 0  2. (a) $a^5$ (b) $b^3$ (c) $c^6$ (d) $d^2$

# 2.2 Expressions

After studying this section, you should be able to understand:

- collecting like terms
- factorising expressions

## Collecting like terms

AQA UNITISED ✓
AQA LINEAR ✓
EDEXCEL A ✓
EDEXCEL B ✓
OCR A ✓
OCR B ✓
WJEC UNITISED ✓
WJEC LINEAR ✓
CCEA ✓

**KEY POINT**

Like terms are those using letters with the same index. Many expressions can be simplified by **collecting like terms**.

**Examples**

Collect like terms:

**(a)** $3b + 2a - c + 5b + 3c$ ← This expression has 1 '$a$' term, 2 '$b$' terms and 2 '$c$' terms

$= 2a + 3b + 5b - c + 3c$ ← Rearrange before simplifying

$= 2a + 8b + 2c$

**(b)** $2x^2 - 3x - x^2 + 5x + y$

$= 2x^2 - x^2 - 3x + 5x + y$ ← You cannot combine $x^2 + 2x$ as they have different indices

$= x^2 + 2x + y$

**(c)** $4pq + p^2 - qp + 2$ ← $q \times p = p \times q = pq$

$= p^2 + 3pq + 2$

Write terms in order of powers.

Multiply out brackets before collecting like terms.

**(d)** $2(x + 1) + 3x(x + 5)$ ← To multiply out brackets, multiply everything inside the brackets by the term outside the bracket. Take care with negative signs.

$= 2 \times x + 2 \times 1 + 3x \times x + 3x \times 5$

$= 2x + 2 + 3x^2 + 15x$

$= 3x^2 + 17x + 2$

## Factorising expressions

AQA UNITISED ✓
AQA LINEAR ✓
EDEXCEL A ✓
EDEXCEL B ✓
OCR A ✓
OCR B ✓
WJEC UNITISED ✓
WJEC LINEAR ✓
CCEA ✓

**KEY POINT**

**Factorising** is the opposite of expanding or multiplying out brackets. Factorising an expression extracts a common factor.

**Examples**

Factorise:

This can be done in one step, but remember each term.

**(a)** $2x^2 - 14x$ ← Common factor 2

$= 2(x^2 - 7x)$ ← Common factor $x$

$= 2x(x - 7)$

**(b)** $2p + 2q + 3mp + 3m^2q$ ← Collect 2 '$p$' terms and 2 '$q$' terms

$= p(2 + 3m) + q(2 + 3m^2)$

> **KEY POINT**
>
> **Quadratic expressions** need to be factorised into two brackets.

To understand how to factorise quadratic expressions you need to know how the brackets were multiplied out.

To multiply out two brackets, the second bracket is multiplied by the first term and then by the second term.

For example:

$(x + a)(x + b)$

$= x \times x + x \times b + a \times x + a \times b$

$= x^2 + bx + ax + ab$

$= x^2 + x(a + b) + ab$

> If $x$ has a coefficient, remember to multiply by this too.

If the signs are different, but the terms are the same the expansion is called the **difference of two squares**.

For example:

$(x + a)(x - a)$

$= x \times x + x \times \text{-}a + a \times x + a \times \text{-}a$

$= x^2 - ax + ax - a^2$

$= x^2 - a^2$

---

**Examples**

Factorise:

**(a)** $y^2 + 5y + 6$

This is a quadratic expression and needs to be factorised into two brackets. The two brackets will look like this:

$(y \quad )(y \quad )$ ⟵ $y \times y$ gives $y^2$

Now look for two factors of 6 which will give 5 when added together. Factors of 6 are $1 \times 6$ and $2 \times 3$. Only $2 + 3 = 5$

$(y + 2)(y + 3)$ ⟵ As both signs in the expression are positive, there must be + signs in both brackets

Check your answer:
Expanding these brackets gives $y^2 + 2y + 3y + 6 = y^2 + 5y + 6$

**(b)** $a^2 + a - 6$ ⟵ Coefficient of $a = 1$

The two brackets will look like this:

$(a \quad )(a \quad )$ ⟵ $a \times a$ gives $a^2$

Factors of 6 are $1 \times 6$ and $2 \times 3$. Only $3 + (\text{-}2) = 1$

$(a + 3)(a - 2)$ ⟵ As the third term in the expression is negative, there must be a + in one bracket and a – in the other bracket

Check your answer:
Expanding these brackets gives $a^2 + 3a - 2a - 6 = a^2 + a - 6$

> If both the second and third terms are positive, both signs in the brackets are positive.
>
> If the second term is negative and the third term is positive, both signs in the brackets are negative.

**(c)** $x^2 - 5x + 6$

The two brackets will look like this:

$(x \quad )(x \quad )$ ← $x \times x$ gives $x^2$

Factors of 6 are $1 \times 6$ and $2 \times 3$. Only $-3 + (-2) = -5$

$(x - 3)(x - 2)$ ← As the third term in the expression is positive, the signs in the brackets could be $+ +$ or $- -$. The second term in the expression is negative, so the signs in the brackets must both be negative.

Check your answer:

Expanding these brackets gives $x^2 - 3x - 2x + 6 = x^2 - 5x + 6$

---

**PROGRESS CHECK**

1. Collect like terms:
   **(a)** $2x + x + y + 3y - 2y$
   **(b)** $2(n + 4) - 3n + m$
   **(c)** $4(a + b) + 2(a - b)$
   **(d)** $3y^2 + xy - y - y^2 + x(y - 1)$
2. Factorise:
   **(a)** $x^2 - 7x$
   **(b)** $y^2 + 5y - 50$
   **(c)** $p^2 + 9p + 20$
   **(d)** $z^2 - z - 12$

1. (a) $3x + 2y$ (b) $m - n + 8$ (c) $6a + 2b$ (d) $2y^2 + 2xy - x - y$ 2. (a) $x(x - 7)$ (b) $(y + 10)(y - 5)$ (c) $(p + 4)(p + 5)$ (d) $(z - 4)(z + 3)$

---

# 2.3 Equations

**LEARNING SUMMARY**

After studying this section, you should be able to understand:

- linear equations
- simultaneous equations
- quadratic equations
- solving simultaneous equations of a line and a circle
- trial and improvement

## Linear equations

| | |
|---|---|
| AQA UNITISED | ✓ |
| AQA LINEAR | ✓ |
| EDEXCEL A | ✓ |
| EDEXCEL B | ✓ |
| OCR A | ✓ |
| OCR B | ✓ |
| WJEC UNITISED | ✓ |
| WJEC LINEAR | ✓ |
| CCEA | ✓ |

**KEY POINT**

A **linear equation** includes an unknown variable with an index of 1.
For example, $2x + 3 = 7$ is a linear equation since $x$ has an index of 1.
When solving linear equations, it is important to keep both sides of the equation balanced at all times.

### Examples

Solve the following equations:

**(a)** $3a = 15$

$$a = \frac{15}{3}$$  ← Divide both sides by 3 so that $a$ is on its own

$$a = 5$$

**(b)** $2x + 3 = 7$

$$2x = 7 - 3$$  ← Subtract 3 from both sides of the equation, so that $2x$ is on its own

$$2x = 4$$

$$x = \frac{4}{2}$$  ← Divide both sides by 2 so that $x$ is on its own

$$x = 2$$

**(c)** $\frac{c}{4} = 3$

$$c = 3 \times 4$$  ← Multiply both sides by 4 so that $c$ is on its own

$$c = 12$$

**(d)** $5x - 20 = 2x - 8$

$$5x - 2x = -8 + 20$$  ← Subtract $2x$ from both sides and add 20 to both sides so that $x$ terms and numbers are separated

$$3x = 12$$

$$x = 4$$

**(e)** $2(x + 5) = 5(x - 7)$  ← Multiply brackets first

$$2x + 10 = 5x - 35$$

$$35 + 10 = 5x - 2x$$

$$45 = 3x$$

$$15 = x$$  ← It does not matter that $x$ is on the right hand side

# Simultaneous equations

| | |
|---|---|
| AQA UNITISED | ✓ |
| AQA LINEAR | ✓ |
| EDEXCEL A | ✓ |
| EDEXCEL B | ✓ |
| OCR A | ✓ |
| OCR B | ✓ |
| WJEC UNITISED | ✓ |
| WJEC LINEAR | ✓ |
| CCEA | ✓ |

**KEY POINT**

**Simultaneous equations** are pairs of equations with two unknown variables. For example, $x + y = 18$ and $x - y = 12$. They are solved using algebra or graphs.

## Solving simultaneous linear equations using algebra

The **substitution** method uses one equation substituted in the other.

### Example

Solve these equations for $x$ and $y$.

$$4x + y = 18 \quad \boxed{1}$$  ← Labelling equations makes them easier to refer to

$$y = 2x \quad \boxed{2}$$

Substitute equation $\boxed{2}$ in equation $\boxed{1}$:  ← Write down your method clearly

$$4x + 2x = 18$$

$$6x = 18$$

$$x = 3$$

From equation $\boxed{2}$ we can see that $y = 2 \times 3 \therefore y = 6$

Check your answer:

LHS (Left Hand Side) of $\boxed{1}$ = $4(3) + 6 = 12 + 6 = 18$

RHS (Right Hand Side) of $\boxed{1}$ = $18$

$\therefore$ the solutions are correct.

$x = 3$ and $y = 6$

> It is a good idea to check your solutions by substituting in equation $\boxed{1}$

The **elimination** method involves manipulating the equations. The first unknown is found by eliminating the second unknown. The first unknown is then substituted in one equation to find the second unknown.

**Examples**

Solve these simultaneous equations:

**(a)**  $x + y = 18$ ①
   $x - y = 12$ ②

Equation ① + equation ② ← $y$ terms have same coefficients so adding will eliminate $y$

$x + x + y + (-y) = 18 + 12$

$2x = 30$ ← $y$ has been eliminated

$x = 15$

Substitute $x = 15$ in equation ① :

$15 + y = 18$

$y = 18 - 15$

$y = 3$

If your checking does not work out, your solutions are incorrect. Go back and check your working.

Check your answer:

LHS of ② $= 15 - 3 = 12$

RHS of ② $= 12$

∴ the solutions are correct.

$x = 15$ and $y = 3$

**(b)**  $4p + q = 17$ ①
   $p + q = 8$ ②

Equation ① − equation ② ← $q$ terms have same coefficients so subtracting will eliminate $q$

$4p - p + q - (+q) = 17 - 8$

$3p = 9$ ← $q$ has been eliminated

$p = 3$

Substitute $p = 3$ in equation ① :

$4(3) + q = 17$

$12 + q = 17$

$q = 17 - 12$

$q = 5$

Check your answer:

LHS of ② $= 3 + 5 = 8$

RHS of ② $= 8$

∴ the solutions are correct.

$p = 3$ and $q = 5$

Remember to multiply all terms on both sides of the equation, so that it remains balanced.

**KEY POINT**

If neither the $x$ nor $y$ terms have the same coefficients, multiply first before using the elimination method.

**Examples**

Solve these simultaneous equations:

(a)
$$x + 3y = 7 \quad \boxed{1}$$
$$2x + 5y = 6 \quad \boxed{2}$$

Equation $\boxed{1} \times 2$ ← *Multiplying by 2 makes the coefficients of $x$ the same*

$$2x + 6y = 14 \quad \boxed{3}$$ ← *This equation needs another label*
$$2x + 5y = 6 \quad \boxed{2}$$

Equation $\boxed{3}$ – equation $\boxed{2}$
$$2x - 2x + 6y - 5y = 14 - 6$$
$$y = 8$$

Substitute $y = 8$ in equation $\boxed{1}$: ← *Use the original equation*
$$x + 3(8) = 7$$
$$x + 24 = 7$$
$$x = 7 - 24$$
$$x = -17$$

Check your answer:

LHS of $\boxed{2}$ = 2(-17) + 5(8) = -34 + 40 = 6

RHS of $\boxed{2}$ = 6

$\therefore$ the solutions are correct.

$x = -17$ and $y = 8$

(b)
$$2x + 5y = 27 \quad \boxed{1}$$
$$3x + 2y = 13 \quad \boxed{2}$$

$\boxed{1} \times 3$ and $\boxed{2} \times 2$ gives $6x$ in both equations

or

$\boxed{1} \times 2$ and $\boxed{2} \times 5$ gives $10y$ in both equations

*None of the coefficients are the same. Equal coefficients cannot be produced by a single multiplication.*

Try the other set of multiplications for yourself.

This example uses $\boxed{1} \times 3$ and $\boxed{2} \times 2$
$$6x + 15y = 81 \quad \boxed{3}$$ ← *These equations need new labels*
$$6x + 4y = 26 \quad \boxed{4}$$

Equation $\boxed{3}$ – equation $\boxed{4}$
$$6x - 6x + 15y - 4y = 81 - 26$$
$$11y = 55$$
$$y = 5$$

Substitute $y = 5$ in equation $\boxed{1}$:
$$2x + 5(5) = 27$$
$$2x + 25 = 27$$
$$2x = 27 - 25$$
$$2x = 2$$
$$x = 1$$

Check your answer:

LHS of $\boxed{2}$ = 3(1) + 2(5) = 3 + 10 = 13

RHS of $\boxed{2}$ = 13

$\therefore$ the solutions are correct.

$x = 1$ and $y = 5$

# Solving simultaneous linear equations using graphs

## Example

Solve these simultaneous equations graphically.

$y = x + 2$

$y = 8 - x$

Draw the lines from the two equations.

See page 67 for how to plot straight-line graphs

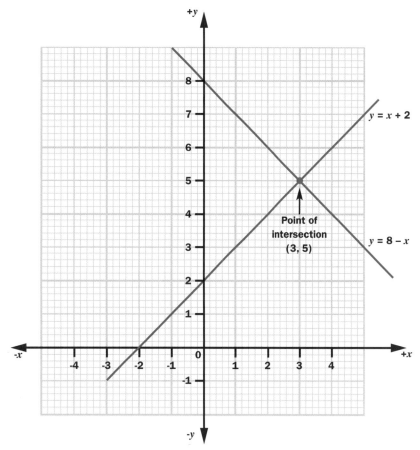

Every point on $y = x + 2$ satisfies that equation.

Every point on $y = 8 - x$ satisfies that equation.

The coordinates of the point of intersection (3, 5) satisfy both equations.

∴ the solution is $x = 3$ and $y = 5$

The coordinates of their point of intersection give the solution of the equations.

# Quadratic equations

AQA UNITISED ✓
AQA LINEAR ✓
EDEXCEL A ✓
EDEXCEL B ✓
OCR A ✓
OCR B ✓
WJEC UNITISED ✓
WJEC LINEAR ✓
CCEA ✓

**KEY POINT**

A **quadratic equation** includes the square of an unknown variable and is generally in the form $ax^2 + bx + c = 0$ although the variable may be a letter other than $x$.

$a$, $b$ and $c$ are constants. As there must always be an $x^2$ term, $a \neq 0$ but $b$ and $c$ can take any values: positive, negative or zero.

There are two solutions to a quadratic equation.

# Solving quadratic equations by factorising

See page 48 for how to factorise expressions.

> **KEY POINT**
>
> To factorise $x^2 + bx + c = 0$ into two brackets $(x + m)(x + n)$, find two numbers $m$ and $n$ where $m + n = b$ and $m \times n = c$.

**Examples**

Solve:

**(a)** $x^2 + 7x + 12 = 0$

$(x + 3)(x + 4) = 0$

> Look for two factors of 12 which give 7 when added together. Factors of 12 are $1 \times 12$, $2 \times 6$ and $3 \times 4$. Only $3 + 4 = 7$

If both signs in the equation are positive, there must be positive signs in both brackets.

When the product of two numbers produces zero, one of the numbers must be zero.

Either $(x + 3) = 0$ or $(x + 4) = 0$

$\therefore \qquad x = -3$ or $x = -4$

**(b)** $x^2 - 7x + 6 = 0$

$(x - 1)(x - 6) = 0$

> Factors of 6 are $1 \times 6$ and $2 \times 3$. Only $-1 + -6 = -7$

If the $b$ term is negative and the $c$ term is positive, the signs in both brackets must be negative.

Either $(x - 1) = 0$ or $(x - 6) = 0$

$\therefore \qquad x = 1$ or $x = 6$

**(c)** $x^2 + x - 12 = 0$

> Coefficient of $x = 1$

$(x + 4)(x - 3) = 0$

> Factors of 12 are $1 \times 12$, $2 \times 6$ and $3 \times 4$. Only $4 + (-3) = 1$

If the $b$ term is positive and the $c$ term is negative, the signs in each bracket must be different.

Either $(x + 4) = 0$ or $(x - 3) = 0$

$\therefore \qquad x = -4$ or $x = 3$

**(d)** $y^2 - 5y - 14 = 0$

$(y - 7)(y + 2) = 0$

> Factors of 14 are $1 \times 14$ and $2 \times 7$. Only $-7 + 2 = -5$

Either $(y - 7) = 0$ or $(y + 2) = 0$

$\therefore \qquad x = 7$ or $x = -2$

**(e)** $2m^2 + 19m + 35 = 0$

> 1st term must be $2m \times m$ as It Is $2m^2$ and both signs must be positive

$(2m + 5)(m + 7) = 0$

> Factors of 35 are $1 \times 35$ and $5 \times 7$; don't forget the $2m$

Either $(2m + 5) = 0$ or $(m + 7) = 0$

$\therefore \qquad m = -2.5$ or $m = -7$

**(f)** $y^2 - 3y = 0$

> In this case $c = 0$

$y(y - 3) = 0$

> Take out common factor $y$

Either $y = 0$ or $(y - 3) = 0$

$\therefore \qquad y = 0$ or $y = 3$

**(g)** $x^2 - 49 = 0$

> In this case $b = 0$

This is the difference of two squares (the signs are different but the terms are the same).

$(x - 7)(x + 7) = 0$

> $\sqrt{49} = \pm 7$

Either $(x - 7) = 0$ or $(x + 7) = 0$

$\therefore \qquad x = 7$ or $x = -7$

## Solving quadratic equations by completing the square

**Completing the square** is a useful method for solving quadratic equations if the equation will not factorise.

> **Example**
>
> Solve $x^2 + 4x = 7$ by completing the square.
>
> $$x^2 + 4x = 7$$ ← Change left hand side of the equation using
> $$(x + 2)^2 - 4 = 7$$ $(x + 2)^2 = (x + 2)(x + 2) = x^2 + 4x + 4$
> $$(x + 2)^2 = 7 + 4$$ ← Add +4 to both sides of the equation
> $$(x + 2)^2 = 11$$
> $$\text{so } x + 2 = \pm\sqrt{11}$$ ← Take the square root of both sides
>
> The two solutions are $x = \text{-}2 + \sqrt{11}$ and $x = \text{-}2 - \sqrt{11}$

> The solutions can be calculated or left as surds.

If the coefficient of $x^2 \neq 1$ then divide first before completing the square.

## Solving quadratic equations by formula

It is often difficult to solve a quadratic equation by factorising or completing the square particularly when $a \neq 1$. In this case use the quadratic formula.

> **KEY POINT**
>
> The solution of the quadratic equation $ax^2 + bx + c = 0$ is given by
> $$x = \frac{\text{-}b \pm \sqrt{b^2 - 4ac}}{2a}$$
> You will be given this formula but you must learn how to use it.

> **Example**
>
> Solve $12x^2 + 7x = 10$ by using the quadratic formula.
>
> $$12x^2 + 7x - 10 = 0$$ ← Rearrange to general form $ax^2 + bx + c = 0$ to find the values of $a, b, c$
> $$a = 12 \quad b = 7 \quad c = \text{-}10$$
> Substitute these values in the quadratic formula:
> $$x = \frac{\text{-}7 \pm \sqrt{7^2 - 4(12 \times \text{-}10)}}{2 \times 12}$$
> $$= \frac{\text{-}7 \pm \sqrt{49 + 480}}{24}$$
> $$= \frac{\text{-}7 \pm \sqrt{529}}{24}$$
> Either $x = \dfrac{\text{-}7 + \sqrt{529}}{24}$ or $x = \dfrac{\text{-}7 - \sqrt{529}}{24}$
> $$\therefore \quad x = 0.67 \text{ (2 d.p.) or } x = \text{-}1.25 \text{ (2 d.p.)}$$

> In the formula $b^2$ must be greater than $4ac$ to obtain a solution

## Solving quadratic equations graphically

Using a graph is not an accurate method of solving a quadratic equation, but it is useful to know how to do it.

> **KEY POINT**
>
> Before solving graphically, a quadratic equation should be in the general form $y = ax^2 + bx + c$
> The curve is drawn and the solutions are where it crosses the $x$-axis ($y = 0$).

See page 71 for how to draw non-linear functions

**Example**

Solve $y = x^2 - x - 1$ graphically. Give solutions to 1 d.p.

| $x$ | -2 | -1 | 0 | 1 | 2 | 3 | 0.5 |
|---|---|---|---|---|---|---|---|
| $y$ | 5 | 1 | -1 | -1 | 1 | 5 | -1.25 |

The coordinates of the minimum point will help with drawing the curve.

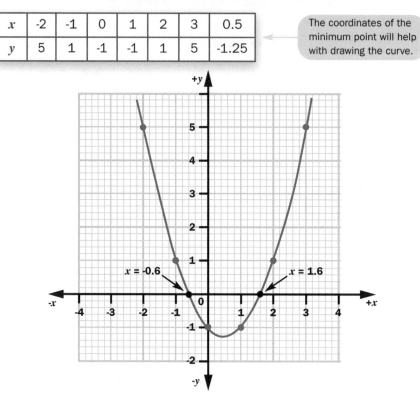

The solutions are where the graph crosses $y = 0$, which is the $x$-axis.

Solutions are 1.6 or -0.6 to 1 d.p.

# Solving simultaneous equations of a line and a circle

| AQA UNITISED | ✓ |
|---|---|
| AQA LINEAR | ✓ |
| EDEXCEL A | ✓ |
| EDEXCEL B | ✓ |
| OCR A | ✓ |
| OCR B | ✓ |
| WJEC UNITISED | ✗ |
| WJEC LINEAR | ✗ |
| CCEA | ✓ |

**KEY POINT**

The solutions of the simultaneous equations of a line and a circle are the points of intersection of the two graphs.

The equation of a circle can be worked out using Pythagoras' theorem. The equation of any circle with centre (0, 0) and radius $r$ is $x^2 + y^2 = r^2$.

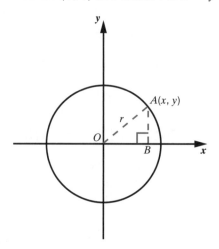

## Example

Solve the equations graphically.

$x^2 + y^2 = 36$ **1**

$x + y = 1$ **2**

Equation **1** is $x^2 + y^2 = 36$ or $x^2 + y^2 = 6^2$, i.e. the equation of a circle of radius 6 units centred at the origin (0, 0).

See page 71 for how to draw non-linear functions

See page 71 for solving other simultaneous equations graphically

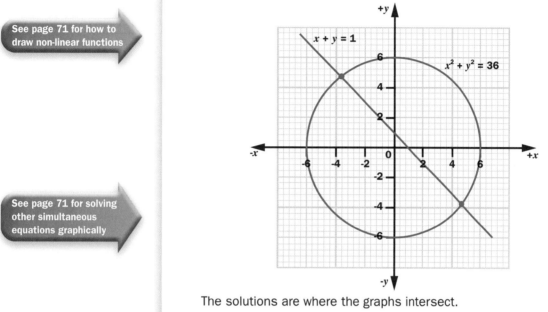

The solutions are where the graphs intersect.

$\therefore x = 4.7$ and $y = -3.7$ or $x = -3.7$ and $y = 4.7$

# Trial and improvement

AQA UNITISED ✓
AQA LINEAR ✓
EDEXCEL A ✓
EDEXCEL B ✓
OCR A ✓
OCR B ✓
WJEC UNITISED ✓
WJEC LINEAR ✓
CCEA ✓

**KEY POINT**

The **trial and improvement** method of solving equations involves testing if a value is near to the solution. Each test is more accurate until the approximate solution is reached.

Read the question carefully to find the degree of accuracy required.

It is sensible to write out your working in a logical way.

## Examples

1. A solution of equation $x^3 + 2x^2 = 200$ lies between 5 and 6. Use trial and improvement to find a solution correct to 1 d.p.

Substitute values of $x$ in LHS of equation and compare with RHS of equation.

| $x$ | $x^3 + 2x^2$ | |
|------|--------------------------------|------------------|
| 5 | 125 + 50 = 175 | Too small |
| 6 | 216 + 72 = 288 | Too large |
| 5.5 | 166.38 + 60.5 = 226.88 | Too large |
| 5.3 | 148.88 + 56.18 = 205.06 | Just too large |
| 5.25 | 144.70 + 55.13 = 199.83 | Just too small |

The value of $x$ that gives exactly 200 is between 5.3 and 5.25

$\therefore x = 5.3$ correct to 1 d.p.

**2.** A solution of equation $x^3 - 8x = 110$ lies between 5 and 6.
Use trial and improvement to find a solution correct to 1 d.p.

| $x$ | $x^3 - 8x$ | |
|---|---|---|
| 5 | $125 - 40 = 85$ | Too small |
| 6 | $216 - 48 = 168$ | Too large |
| 5.5 | $166.38 - 44 = 122.38$ | Too large |
| 5.3 | $148.88 - 42.4 = 106.48$ | Just too small |
| 5.35 | $153.13 - 42.8 = 110.33$ | Just too large |

The solution is between 5 and 5.5, so a smaller increment is tried.

The value of $x$ that gives exactly 100 is between 5.3 and 5.35
$\therefore x = 5.3$ correct to 1 d.p.

### PROGRESS CHECK

1. Solve these linear equations:

   **(a)** $2(2x + 3) = 42$  **(b)** $\dfrac{y}{2} - 4 = 1$

   **(c)** $3p + 4 = 2 - p$  **(d)** $23 = 9 - 4m$

2. Solve these simultaneous equations algebraically:
   **(a)** $a = 8 - 2b$  $\quad 2a + b = 7$
   **(b)** $3m - 2n = 4$  $\quad m + 4n = 6$

3. Solve these simultaneous equations graphically:
   **(a)** $2p - q = 2$  $\quad q = 7 - p$
   **(b)** $a = 7 - 4b$  $\quad 2a + 3b = 9$

4. Solve these quadratic equations by factorising:
   **(a)** $p^2 - 9p + 14 = 0$
   **(b)** $3n^2 + 13n + 4 = 0$

5. Solve these quadratic equations by completing the square:
   **(a)** $y^2 + 3y - 10 = 0$
   **(b)** $3x^2 + 36x + 12 = 0$

6. Solve these quadratic equations by using the quadratic formula giving
   the answer to 2 d.p.
   **(a)** $2x^2 + 6x + 3 = 0$
   **(b)** $y^2 + 4y - 2 = 0$

1. (a) $x = 9$ (b) $y = 10$ (c) $p = -\frac{1}{2}$ (d) $m = -3\frac{1}{2}$  2. (a) $a = 2$ and $b = 3$ (b) $m = 2$ and $n = 1$
3. (a) $p = 3$ and $q = 4$ (b) $a = 3$ and $b = 1$  4. (a) $p = 7$ or 2 (b) $n = -\frac{1}{3}$ or $-4$
5. (a) $y = 2$ and $-5$ (b) $x = -6 \mp 4\sqrt{2}$  6. (a) $x = -0.63$ or $-2.37$ (b) $y = 0.45$ or $y = -4.45$

# 2.4 Formulae

> **LEARNING SUMMARY**
>
> **After studying this section, you should be able to understand:**
>
> - deriving a formula
> - substituting in a formula
> - changing the subject of a formula

## Deriving a formula

AQA UNITISED ✓
AQA LINEAR ✓
EDEXCEL A ✓
EDEXCEL B ✓
OCR A ✓
OCR B ✓
WJEC UNITISED ✓
WJEC LINEAR ✓
CCEA ✓

> **KEY POINT**
>
> A formula is an equation that shows a relationship between two or more quantities. For example:
>
> $A = \frac{1}{2}bh$      $C = \pi d$      $v = u + at$

You may be asked to derive a formula representing a relationship.

*Think of deriving a formula as translating given information into mathematical language. If you need to use your own letter symbols, explain what they mean.*

### Examples

1. The length of a rectangle is $4x$ and the width is $y$. Write a formula representing the perimeter ($P$).

   Perimeter = 2 × (length + width)
   $P = 2(4x + y)$

2. Esther delivers newspapers. She is paid £10 per week, but $x$ pence is deducted for each incorrectly delivered newspaper. In her first two weeks, five papers are delivered per week to the wrong houses. After two weeks, she is paid £18.

   **(a)** Write an expression for her pay for the first week.

   **(b)** How much does she lose for each misdelivered paper?

   **(a)** Let pay for 1st week be $P$.
   $P = 1000 - 5x$ pence        ← Make sure all terms are in same units

   **(b)** Pay for 2 weeks:
   $2(1000 - 5x) = 1800$
   $2000 - 10x = 1800$
   $1000 - 5x = 900$        ← Divide both sides by 2
   $1000 - 900 = 5x$        ← Subtract 900 from both sides and add $5x$ to both sides
   $100 = 5x$
   $20 = x$

   Esther has 20p deducted from her wages for each misdelivered paper.

*Always give the answer in terms of the original question.*

## Substituting in a formula

AQA UNITISED ✓
AQA LINEAR ✓
EDEXCEL A ✓
EDEXCEL B ✓
OCR A ✓
OCR B ✓
WJEC UNITISED ✓
WJEC LINEAR ✓
CCEA ✓

**KEY POINT**

**Substituting** given values for some variables in a formula means you can work out other values.

> Method marks are often given if the values are seen to have been substituted correctly. Do not try to jump this step.

**Examples**

**1.** The area of a triangle is given by the formula $A = \frac{1}{2}bh$.

Find $A$ when $b = 7$cm and $h = 6$cm.

$A = \frac{1}{2} \times 7 \times 6$ — $A$ is the subject of this formula. Replace each letter with the given value

$A = 7 \times 3$ — Cancel by 2

$A = 21$cm$^2$ — Area is measured in cm$^2$

**2.** Temperature is converted from °F to °C by the formula $C = \frac{5(F - 32)}{9}$

What is 75°F in °C? Give your answer to the nearest whole number.

$C = \frac{5(75 - 32)}{9}$

$C = \frac{5 \times 43}{9}$

$C = 23.\dot{8}$

$C = 24$°C to nearest degree

> If you are substituting negative numbers, take care with signs.

## Changing the subject of a formula

AQA UNITISED ✓
AQA LINEAR ✓
EDEXCEL A ✓
EDEXCEL B ✓
OCR A ✓
OCR B ✓
WJEC UNITISED ✓
WJEC LINEAR ✓
CCEA ✓

You may need to rearrange a formula to make a calculation easier.

**KEY POINT**

When **changing the subject of a formula**, follow the same rules for manipulating an equation.

**Example**

Find 18°C in °F using the temperature formula $C = \frac{5(F - 32)}{9}$

You can substitute 18°C into the formula and then calculate °F or you can change the subject of the formula to F.

$C = \frac{5(F - 32)}{9}$

$9C = 5(F - 32)$ — Multiply both sides by 9

$\frac{9C}{5} = F - 32$ — Divide both sides by 5

$\frac{9C}{5} + 32 = F$ — Add 32 to both sides

> This can be written $F = \frac{9C}{5} + 32$

Now substitute C = 18° into the rearranged formula.

$F = \frac{9 \times 18}{5} + 32 = 32.4 + 32$

$\therefore F = 64.4$°F

**1** **(a)** A cooking guide for roasting lamb is given as 30mins per given weight ($w$) plus an extra 20mins. Write a formula to find how long ($T$) it would take to roast a shoulder of lamb.

**(b)** The cost ($C$) of hiring a car, with mileage included, is £116 per week ($w$) plus £2.20 insurance per day ($d$). Write a formula to find $C$.

**2** Calculate by substituting the given values in these formulae. Give answers to 3 s.f. where necessary.

**(a)** $V = lbh$        [$l = 6$, $b = 5.5$, $h = 3$]

**(b)** $v = u + at$        [$a = 2.5$, $u = 4.3$, $t = 4$]

**(c)** $A = \pi r^2$        [$\pi = 3.142$, $r = 7$]

**(d)** $v^2 = u^2 + 2as$        [$u = 8$, $a = -3$, $s = 4.2$]

**3** Make the letter given in brackets the subject of these formulae:

**(a)** $I = \dfrac{PRT}{100}$        ($R$)

**(b)** $v^2 = u^2 + 2as$        ($u$)

**(c)** $C = 2\pi r$        ($r$)

**(d)** $T = \sqrt{\dfrac{p}{q}}$        ($p$)

3. (a) $R = \dfrac{100I}{PT}$ (b) $u = \sqrt{v^2 - 2as}$ (c) $r = \dfrac{C}{2\pi}$ (d) $p = T^2 q$

1. (a) $T = 30w + 20$ (b) $C = 116w + 2.20d$ 2. (a) 99 (b) 14.3 (c) 154 to 3 s.f. (d) 6.23 to 3 s.f.

# 2.5 Inequalities

| LEARNING SUMMARY | **After studying this section, you should be able to understand:** |
|---|---|
| | • representing inequalities |
| | • solving an inequality |

# Representing inequalities

| AQA UNITISED | ✓ |
|---|---|
| AQA LINEAR | ✓ |
| EDEXCEL A | ✓ |
| EDEXCEL B | ✓ |
| OCR A | ✓ |
| OCR B | ✓ |
| WJEC UNITISED | ✓ |
| WJEC LINEAR | ✓ |
| CCEA | ✓ |

These symbols are used to **represent inequalities**:

- $x < 4$    ($x$ is less than 4)
- $x > 7$    ($x$ is greater than 7)
- $y \leqslant 10$    ($y$ is less than or equal to 10)
- $y \geqslant -3$    ($y$ is greater than or equal to -3)

Number lines are also used to illustrate inequalities:

- An included number is shown by a filled circle.
- A number not included is shown by an empty circle.

## Examples

Show the following on number lines.

**(a)** $x \geqslant -2$

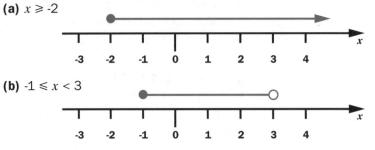

**(b)** $-1 \leqslant x < 3$

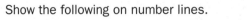

**(c)** $5 \geqslant 2y - 1 > 1$

This must be arranged so that $y$ is on its own. The inequality is a combination of:

$5 \geqslant 2y - 1$ and $2y - 1 > 1$

$6 \geqslant 2y$       $2y > 2$   ←   Add 1 to both sides in both inequalities

$3 \geqslant y$        $y > 1$   ←   Divide both sides by 2

so $3 \geqslant y > 1$

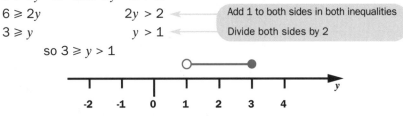

# Solving an inequality

| | |
|---|---|
| AQA UNITISED | ✓ |
| AQA LINEAR | ✓ |
| EDEXCEL A | ✓ |
| EDEXCEL B | ✓ |
| OCR A | ✓ |
| OCR B | ✓ |
| WJEC UNITISED | ✓ |
| WJEC LINEAR | ✓ |
| CCEA | ✓ |

**KEY POINT**

When **solving an inequality**, follow the same procedure used for solving an equation, except when multiplying or dividing each side by a negative number you must reverse the inequality.

## Solving an inequality with one variable

### Examples

Solve these inequalities:

**(a)** $5 - 3x \leqslant 1$

$5 \leqslant 1 + 3x$   ←   Add $3x$ to both sides

$5 - 1 \leqslant 3x$   ←   Subtract 1 from both sides

$4 \leqslant 3x$

$\dfrac{4}{3} \leqslant x$   ←   Divide both sides by 3

or $x \geqslant 1\frac{1}{3}$

**(b)** $8 - 2x > 10$

It is sensible to divide the whole inequality by 2 so that the coefficient of $x = 1$.

$4 - x > 5$

$-x > 5 - 4$   ←   Subtract 4 from both sides

$-x > 1$

Now divide the whole inequality by -1, so that you will have $x$ on LHS instead of $-x$.

Remember that dividing by a negative number reverses the inequality.

$\therefore x < -1$

## Solving an inequality with two variables

For the equation of a straight-line graph, see page 67.

> **KEY POINT**
>
> If an inequality has two variables such as $x + y > 4$ you need to draw the straight line of the corresponding equation. Look at the areas above and below the line. These are called regions and are the solutions of the inequality.

### Example

Solve $x + y > 4$ and shade the region that satisfies this inequality.

Rearrange the equation into the form $y = mx + c$. This gives $y = -x + 4$. Work out the coordinates for $y = -x + 4$ and draw a straight-line graph.

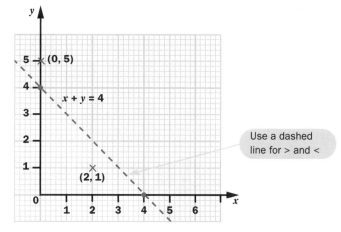

Use a dashed line for > and <

The line $y = 4 - x$ intercepts both axes at 4. Try substituting $x = 0$, then $y = 0$ to see why.

Choose two points, one below the line and one above, then substitute coordinates into LHS of inequality.

(2, 1) below the line gives $2 + 1 = 3 \rightarrow x + y < 4$

(0, 5) above the line gives $0 + 5 = 5 \rightarrow x + y > 4$

All the points below the line give $x + y < 4$

so the shaded area above the line gives $x + y > 4$

## Graphs showing more than one inequality

### Example

Shade the region that is satisfied by $x \geqslant 0$, $y \leqslant 2$ and $4y + 3x < 12$

The equation of the $x$-axis is $y = 0$.

The equation of the $y$-axis is $x = 0$.

$y = 2$ at every point on this line

$x = 0$ at every point on this line

Use a solid line for $\leqslant$ and $\geqslant$

$y = 2$

$x = 0$

$4y + 3x = 12$

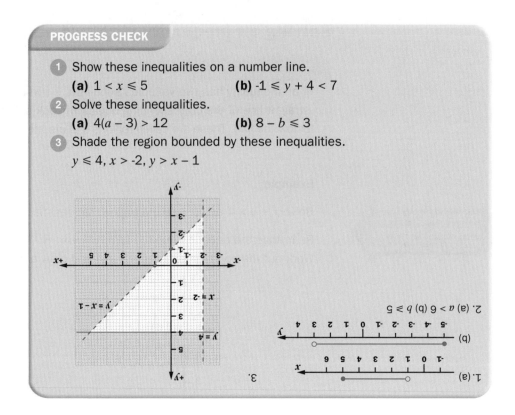

# 2.6 Sequences

| LEARNING SUMMARY | After studying this section, you should be able to understand: |
|---|---|
| | • generating sequences |
| | • $n$th term |

## Generating sequences

| | |
|---|---|
| AQA UNITISED | ✓ |
| AQA LINEAR | ✓ |
| EDEXCEL A | ✓ |
| EDEXCEL B | ✓ |
| OCR A | ✓ |
| OCR B | ✓ |
| WJEC UNITISED | ✓ |
| WJEC LINEAR | ✓ |
| CCEA | ✓ |

**KEY POINT**

A **sequence** is a collection of **terms** that follow a pattern or rule.

Sequences can be generated in several ways. You may be given the first few terms of a sequence and asked to generate the next few terms.

There are many other possible patterns. You may have to multiply a term and then add a number to get the next term. It is important to find the pattern or rule governing the sequence.

- Some sequences are generated by adding or subtracting the same number each time, called a **common difference**. For example:

  **1, 4, 7, 10, ...**   ←   Each term is found by adding 3 to the previous term
  +3   +3   +3

- Some sequences are generated by multiplying or dividing by the same number, called a **constant**. For example:

  **2, 4, 6, 8, ...**   ←   The position of the term is multiplied by 2
  $1{\times}2$   $2{\times}2$   $3{\times}2$   $4{\times}2$

- Some sequences are square numbers or cube numbers. For example:

  **1, 4, 9, 16, ...**   ←   The position of the term is squared
  $1^2$   $2^2$   $3^2$   $4^2$

# *n*th term

**KEY POINT**

The ***n*th term** is a general formula for a sequence where *n* is the position of the term, i.e. the first term has $n = 1$.

Watch out for squares and cubes of numbers as well as triangular numbers.

**Example**

Find the *n*th term of these sequences.

**(a)** 2, 4, 6, 8, …

1st term $= 1 \times 2 = 2$ ← It is a good idea to number the terms
2nd term $= 2 \times 2 = 4$
3rd term $= 3 \times 2 = 6$
4th term $= 4 \times 2 = 8$

Position of term    Constant    Term

*n*th term $= n \times 2$
$= 2n$

**(b) (i)** 5, 8, 11, 14, …

1st term $= 1 \times 3 + 2 = 5$ ← The constant is 3, but the first term is 5 so you also need to add 2
2nd term $= 2 \times 3 + 2 = 8$
3rd term $= 3 \times 3 + 2 = 11$
4th term $= 4 \times 3 + 2 = 14$

Position of term    × constant + 2    Term

*n*th term $= n \times 3 + 2$
$= 3n + 2$

**(ii)** If the sequence had been written 2, 5, 8, 11, …

1st term $= 0 \times 3 + 2 = 2$
2nd term $= 1 \times 3 + 2 = 5$ ← The constant 3 is multiplied by 1 less than the position of the term
3rd term $= 2 \times 3 + 2 = 8$
4th term $= 3 \times 3 + 2 = 11$

*n*th term $= 3(n - 1) + 2$
$= 3n - 3 + 2$
$= 3n - 1$

You may be asked to find specific terms of a sequence by substituting in the *n*th term.

**Example**

Find the 8th term of a sequence with *n*th term $= 2n - 3$.

8th term $= 2(8) - 3$
$= 16 - 3$
$= 13$

# 2.7 Functions

| | After studying this section, you should be able to understand: |
|---|---|
| **LEARNING SUMMARY** | • coordinates<br>• linear functions<br>• graphs for real-life situations<br>• non-linear functions<br>• transformation of functions |

## Coordinates

AQA UNITISED ✓
AQA LINEAR ✓
EDEXCEL A ✓
EDEXCEL B ✓
OCR A ✓
OCR B ✓
WJEC UNITISED ✓
WJEC LINEAR ✓
CCEA ✓

### KEY POINT

**Coordinates** are pairs of numbers that give the position of a point on a graph or grid. They are given in the form $(x, y)$ and plotted using a dot or small cross.

> A sharp pencil is necessary to plot and join the points as accurately as possible.

> Label the axes as shown. Make sure there is a comma between coordinates and give the $x$ coordinate first.

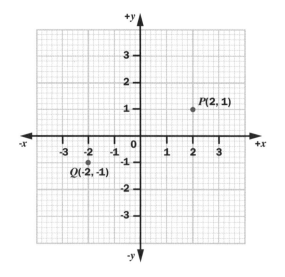

# Linear functions

A straight-line graph can be drawn using two points, but it is more accurate to use at least three points.

A linear function is given in the form:

$y = mx + c$ ← c is the intercept on the y-axis

m is the gradient

$y = mx + c$ is the general equation of a straight line and is the rule for all points on the line. To draw a straight-line graph a selection of coordinates is plotted. It is useful to have a table of coordinates. When you are given values of $x$, the values of $y$ can be calculated. You may be given a range of values or have to choose your own.

Remember signs when calculating.

---

**Example**

Draw the graph of $y = 2x + 1$ for $-2 \leqslant x \leqslant 1$.

| x | -2 | -1 | 0 | 1 |
|---|----|----|---|---|
| y | -3 | -1 | 1 | 3 |

Substitute values of $x$ in $2x + 1$ to find values of $y$

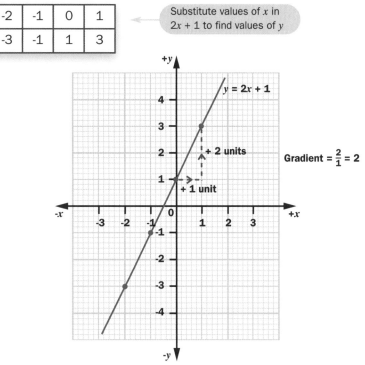

Gradient = $\frac{2}{1}$ = 2

The line cuts the $y$-axis at $y = 1$. This is called the **intercept**.

The slope of the line is the **gradient**. This is calculated by:

$$\frac{\text{Vertical change}}{\text{Horizontal change}}$$

If the line goes 'uphill', as in this example, the gradient is positive. If the line goes 'downhill' the gradient is negative. As the gradient increases, the line gets steeper.

---

If the intercept is 0, the line passes through the origin (0, 0).

## Finding the equation of a straight-line graph

The equation of a straight-line graph can be found if you are given the gradient and the $y$-intercept.

**Examples**

1. What are the equations of these straight lines?

   **(a)** Gradient = 1, passing through point (0, -5)

   $$y = mx + c$$ ⟵ General equation
   $$y = (1)x + (-5)$$
   $$y = x - 5$$

   **(b)** Gradient = $-\frac{3}{4}$, passing through point (0, 2)

   $$y = mx + c$$
   $$y = (-\frac{3}{4})x + (2)$$
   $$y = -\frac{3}{4}x + 2$$

   This could be written as $4y = 8 - 3x$ ⟵ Multiply both sides by 4

2. Find the gradient and $y$-intercept of this straight line.

   $$y = 4x + 1$$ ⟵ Compare with $y = mx + c$
   gradient = 4 as $m = 4$
   $y$-intercept = 1 as $c = 1$

## Parallel lines

**KEY POINT**

If lines are parallel, they have the same gradient.

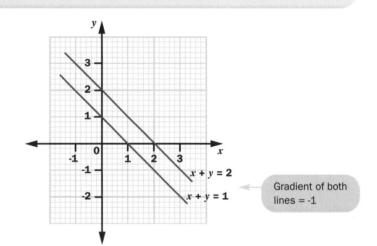

$x + y = 2$

$x + y = 1$

Gradient of both lines = -1

**Example**

What is the equation of a line passing through (0, -2) and parallel to $y = 3x + 2$?

Gradient must be 3 as $m = 3$ and lines are parallel.
$y$-intercept = -2
∴ equation is $y = 3x - 2$

If the gradients ($m$) are equal, lines are parallel.

## Perpendicular lines

**KEY POINT**

If lines are perpendicular to each other, i.e. at 90° to one another, the product of their gradients = -1.

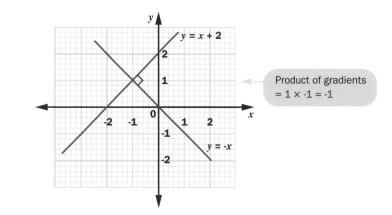

Product of gradients
= 1 × -1 = -1

### Example

What is the gradient of a line perpendicular to $y = 3x + 1$?

If gradient of first line is $m_1$, then $m_1 = 3$.

If gradient of perpendicular line is $m_2$

$m_1 \times m_2 = -1$

$3 \times m_2 = -1$

$3m_2 = -1$

$m_2 = -\dfrac{1}{3}$ ⟵ Gradient of perpendicular line

If the product of gradients $(m_1 m_2) = -1$, lines are perpendicular to each other.

# Graphs for real-life situations

AQA UNITISED ✓
AQA LINEAR ✓
EDEXCEL A ✓
EDEXCEL B ✓
OCR A ✓
OCR B ✓
WJEC UNITISED ✓
WJEC LINEAR ✓
CCEA ✓

Information for real-life situations can be represented on graphs. One quantity is plotted against another quantity, such as distance and time.

● **Distance–time graph**
  Joseph's journey is shown on this graph.

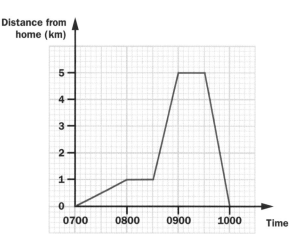

Time is always on the horizontal axis. Read values for time and distance off the graph.

The gradient of each sloping line is Joseph's speed at that particular time.

0700 – 0800: speed = 1km ÷ 1hr
            = 1km/hr

0800 – 0830: the line is horizontal (gradient = 0)
This means Joseph is stationary.

0830 – 0900: speed = 4km ÷ 0.5hr
            = 8km/hr

0900 – 0930: Joseph is stationary.

0930 – 1000: Joseph returns home at speed = 5km ÷ 0.5hr = 10km/hr

The slope is steeper between 0930 and 1000 indicating a faster speed.

- **Speed–time graphs**

  The gradient of a speed–time graph shows acceleration. The steeper the slope, the greater the acceleration. A horizontal line means a steady speed with no acceleration.

  It may look like this:

If WJEC is your exam board, you may be asked to find the area under a velocity-time graph. This is the distance travelled.

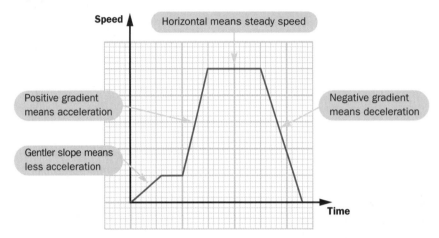

- **Matching graphs to situations**

  Three containers are filled with water. These graphs illustrate how the depth of water changes with time.

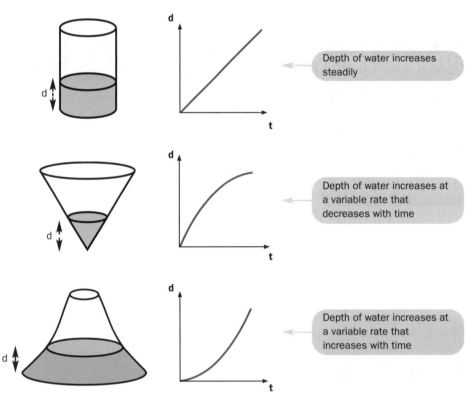

# Non-linear functions

| | |
|---|---|
| AQA UNITISED | ✓ |
| AQA LINEAR | ✓ |
| EDEXCEL A | ✓ |
| EDEXCEL B | ✓ |
| OCR A | ✓ |
| OCR B | ✓ |
| WJEC UNITISED | ✓ |
| WJEC LINEAR | ✓ |
| CCEA | ✓ |

To draw a graph of any non-linear function, calculate a table of coordinates and plot these points. The result will be a curve, not a straight line, so do not use a ruler to join points. Draw a smooth curve through all the points, continuing past the final points plotted. If a point seems out of line, check your calculation.

## Quadratic function

> **KEY POINT**
>
> A **quadratic function** will include the square of a variable. The curve will always be U-shaped and is also called a parabola.

**Examples**

1. Draw the graph of $y = x^2 + 2$

| $x$ | -2 | -1 | 0 | 1 | 2 |
|---|---|---|---|---|---|
| $y$ | 6 | 3 | 2 | 3 | 6 |

*+x and -x give the same values of y as $x^2$ is used*

Use a sharp pencil to draw the graph. Do not use a pen as you may make a mistake.

Marks are awarded for drawing a smooth curve.

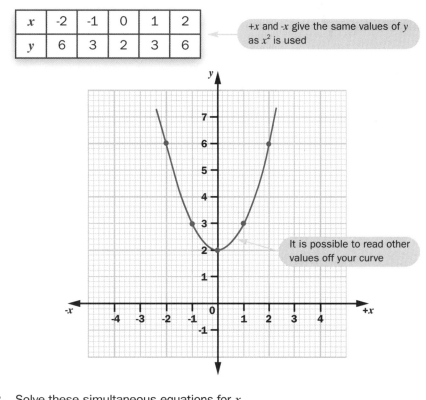

*It is possible to read other values off your curve*

2. Solve these simultaneous equations for $x$.

$y = x^2 + 3$
$y = 4 - x$

*The point of intersection will give the solutions as the coordinates will satisfy both equations*

Draw the graphs for $-2 \leqslant x \leqslant 2$

$y = x^2 + 3$

| $x$ | -2 | -1 | 0 | 1 | 2 |
|---|---|---|---|---|---|
| $y$ | 7 | 4 | 3 | 4 | 7 |

$y = 4 - x$

| $x$ | -2 | -1 | 0 | 1 | 2 |
|---|---|---|---|---|---|
| $y$ | 6 | 5 | 4 | 3 | 2 |

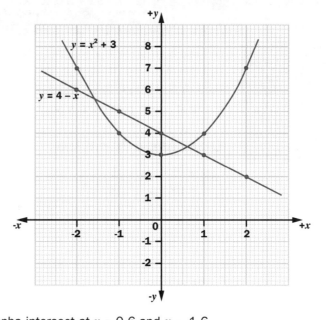

You do not need to give the $y$ coordinate in this example as you are solving the equations for $x$

The graphs intersect at $x = 0.6$ and $x = -1.6$

## Cubic function

> **KEY POINT**
>
> A **cubic function** will include the cube of a variable. It will always be S-shaped.

**Example**

Draw the graph of $y = x^3 + x - 5$

| $x$ | -2 | -1 | 0 | 1 | 2 |
|-----|-----|-----|-----|-----|-----|
| $y$ | -15 | -7 | -5 | -3 | 5 |

When you substitute negative values in $x^3$, the result is negative

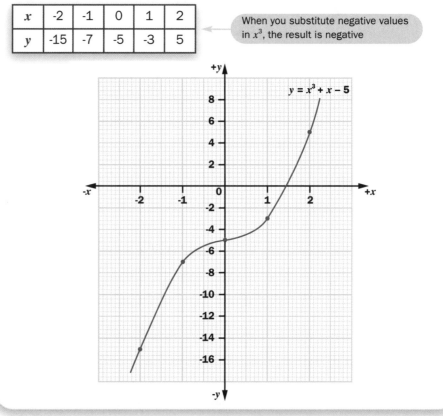

## Reciprocal function

A **reciprocal function** always has $x$ in the denominator of the fraction:

$$f(x) = \frac{1}{x} \qquad y = \frac{2}{x} \qquad f(x) = \frac{1}{3 + x}$$

### Example

Draw the graph of $y = \dfrac{3}{x}$ for $-4 \leqslant x \leqslant 4$.

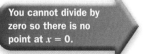

You cannot divide by zero so there is no point at $x = 0$.

| $x$ | -4 | -3 | -2 | -1 | 1 | 2 | 3 | 4 |
|-----|----|----|----|----|----|----|----|----|
| $y$ | $-\frac{3}{4}$ | -1 | $-1\frac{1}{2}$ | -3 | 3 | $1\frac{1}{2}$ | 1 | $\frac{3}{4}$ |

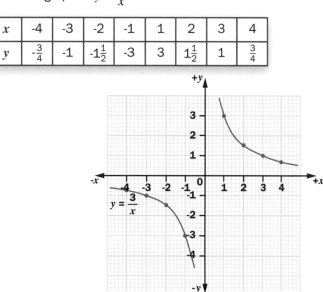

## Exponential function

See page 13 for index notation.

The **exponential function** is in the form $f(x) = k^x$ where $k$ is a constant:

$$y = 2^x \qquad f(x) = 3^x$$

The index or exponent is the variable $x$.

### Example

Draw the graph of $y = 2^x$ for $-4 \leqslant x < 3$.
Give values of $y$ to 2 d.p.

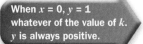

When $x = 0$, $y = 1$ whatever of the value of $k$. $y$ is always positive.

| $x$ | -4 | -3 | -2 | -1 | 0 | 1 | 2 |
|-----|----|----|----|----|----|----|----|
| $y$ | 0.06 | 0.13 | 0.25 | 0.5 | 1 | 2 | 4 |

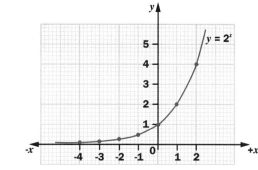

See pages 93–98 for trigonometry.

# Trigonometrical functions

> **KEY POINT**
>
> You need to be able to recognise the graphs of the following **trigonometric functions**:
>
> $f(x) = \sin x°$     $f(x) = \cos x°$     $f(x) = \tan x°$
>
> These are also called **circular functions**.
>
> They are periodic functions as their values repeat. The sine and cosine functions have a period of 360° and the tangent function has a period of 180°.

The graphs $y = \sin x°$, $y = \cos x°$ and $y = \tan x°$ are plotted for $0 \leqslant x \leqslant 360°$

$y$ values are found using a calculator.

● **Sine function:** $y = \sin x°$

The sine curves lie between 1 and -1.

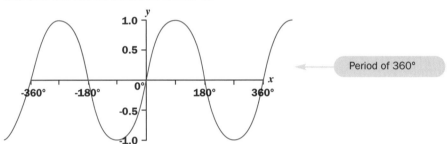

Period of 360°

● **Cosine function:** $y = \cos x°$

The cosine curves lie between 1 and -1.

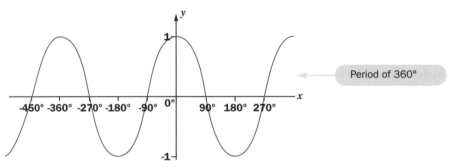

Period of 360°

● **Tangent function:** $y = \tan x°$

There is no limit on the tangent curves.

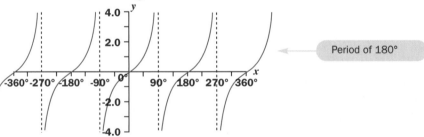

Period of 180°

# Transformation of functions

| AQA UNITISED | ✓ |
| AQA LINEAR | ✓ |
| EDEXCEL A | ✓ |
| EDEXCEL B | ✓ |
| OCR A | ✓ |
| OCR B | ✓ |
| WJEC UNITISED | ✓ |
| WJEC LINEAR | ✓ |
| CCEA | ✗ |

The graph of a function can be transformed by...

● translating

● stretching

● compressing

● reflecting.

Both the position and size of the graph may be affected.

## Translation of functions

If $y$ = 'an expression involving $x$' then it can be written as $y = f(x)$.

If $f(x)$ is the function and $a > 0$
- $y = f(x) + a$ moves the original graph '$a$' units up ($\uparrow$) the $y$-axis
- $y = f(x) - a$ moves the original graph '$a$' units down ($\downarrow$) the $y$-axis
- $y = f(x - a)$ moves the original graph '$a$' units to the right ($\rightarrow$)
- $y = f(x + a)$ moves the original graph '$a$' units to the left ($\leftarrow$)

> This moves the graph in the opposite direction to what you would expect.

For example:

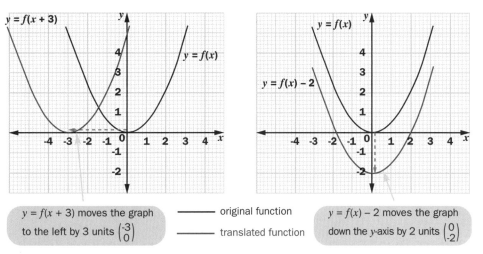

$y = f(x + 3)$ moves the graph to the left by 3 units $\begin{pmatrix} -3 \\ 0 \end{pmatrix}$

—— original function

—— translated function

$y = f(x) - 2$ moves the graph down the $y$-axis by 2 units $\begin{pmatrix} 0 \\ -2 \end{pmatrix}$

## Stretching and compressing functions

If $f(x)$ is the function
- $y = af(x)$ stretches the original graph along the $y$-axis by a factor of $a$. If $a > 1$, e.g. $y = 2x^2$, the $y$-coordinates of all the points are multiplied by a factor of 2. If $a < 1$, e.g. $y = 0.5x^2$, the $y$-coordinates are multiplied by a factor of 0.5
- $y = f(ax)$ stretches the original graph along the $x$-axis by a factor of $a$. If $a > 1$, all the points are stretched inwards in the $x$ direction by a factor of $\frac{1}{a}$, e.g. if $y = (2x)^2$, the $x$-coordinates are multiplied by $\frac{1}{a} = \frac{1}{2} = 0.5$. If $a < 1$, all the points are stretched outwards in the $x$ direction by a factor of $\frac{1}{a}$, e.g. if $y = (0.5x)^2$, the $x$-coordinates are multiplied by $\frac{1}{a} = \frac{1}{0.5} = 2$

For example:

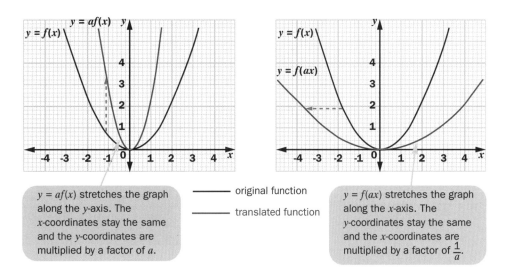

$y = af(x)$ stretches the graph along the $y$-axis. The $x$-coordinates stay the same and the $y$-coordinates are multiplied by a factor of $a$.

—— original function

—— translated function

$y = f(ax)$ stretches the graph along the $x$-axis. The $y$-coordinates stay the same and the $x$-coordinates are multiplied by a factor of $\frac{1}{a}$.

# Reflection of functions

See page 105
Transformations.

> **KEY POINT**
>
> Reflection of function $y = f(x)$ in the $x$-axis gives $y = -f(x)$.
>
> Reflection of function $y = f(x)$ in the $y$-axis gives $y = f(-x)$.

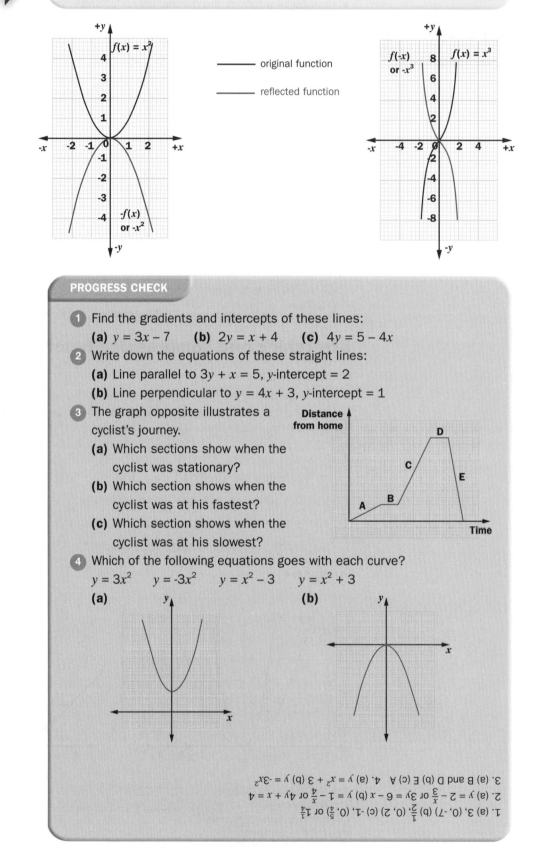

— original function

— reflected function

> **PROGRESS CHECK**
>
> 1. Find the gradients and intercepts of these lines:
>    **(a)** $y = 3x - 7$ **(b)** $2y = x + 4$ **(c)** $4y = 5 - 4x$
> 2. Write down the equations of these straight lines:
>    **(a)** Line parallel to $3y + x = 5$, $y$-intercept $= 2$
>    **(b)** Line perpendicular to $y = 4x + 3$, $y$-intercept $= 1$
> 3. The graph opposite illustrates a cyclist's journey.
>    **(a)** Which sections show when the cyclist was stationary?
>    **(b)** Which section shows when the cyclist was at his fastest?
>    **(c)** Which section shows when the cyclist was at his slowest?
> 4. Which of the following equations goes with each curve?
>    $y = 3x^2$ $y = -3x^2$ $y = x^2 - 3$ $y = x^2 + 3$
>    **(a)** **(b)**

**Distance from home** / **Time** (A, B, C, D, E)

# Sample GCSE questions

**1** A carpet is sold in rolls of 50 metres in length and 3 metres width.

   **(a)** A customer buys $m$ metres of the carpet. What is the area of carpet bought? **(1)**

   **(b)** What length of carpet is left on the roll? **(1)**

   **(c)** The customer wishes to buy more pieces of carpet of the same length. How many can be cut from the remaining length? **(1)**

**Area = length × width** →
(a) Area $= 3 \times m = 3m\,\text{m}^2$

(b) Length $= 50 - m$ metres

**Divide the remainder by the piece length** →
(c) Number of lengths $= \dfrac{50 - m}{m}$

**2** Solve these equations:

   **(a)** $5x^2 - 4x - 1 = 0$ (Use the quadratic formula) **(5)**

   **(b)** $2y^2 - 6y + 1 = 0$ (Use completing the square) **(5)**

(a) $x = \dfrac{-b \pm \sqrt{b^2 - 4ac}}{2a} = \dfrac{-(-4) \pm \sqrt{(-4)^2 - 4(5 \times -1)}}{2 \times 5}$

**Add the square root. Never split before doing square root** →
$= \dfrac{+4 \pm \sqrt{16 + 20}}{10} = \dfrac{+4 \pm \sqrt{36}}{10} = \dfrac{+4 \pm 6}{10}$

**There are two solutions using ±** →
Either $x = \dfrac{(4 + 6)}{10}$ or $x = \dfrac{(4 - 6)}{10}$

$\therefore x = 1$ or $-\dfrac{1}{5}$

**Divide by 2 so that the coefficient of $y^2$ = 1** →
(b) $2y^2 - 6y + 1 = 0 \longrightarrow y^2 - 3y + \dfrac{1}{2} = 0$

**Divide the coefficient of $y$ by 2 to give $\frac{3}{2}$** →
$(y - \dfrac{3}{2})^2 = y^2 - 3y + \dfrac{9}{4}$

Comparing with original equation gives $y^2 - 3y + \dfrac{9}{4} - \dfrac{7}{4} = 0$

**Square root both sides; $\sqrt{4}$ = 2** →
Rearrange to $y^2 - 3y + \dfrac{9}{4} = \dfrac{7}{4}$ or $(y - \dfrac{3}{2})^2 = \dfrac{7}{4} \longrightarrow y - \dfrac{3}{2} = \pm\dfrac{\sqrt{7}}{2}$

**Multiply by 2 to eliminate the denominator** →
Either $2y = 3 + \sqrt{7}$ or $2y = 3 - \sqrt{7}$

$y = 2.8$ or $0.2$ (1 d.p.)

**3** Factorise these expressions:

   **(a)** $8p^3 + 6pqr$ **(2)**

   **(b)** $m^2 - 4m - 32$ **(2)**

   **(c)** $9x^2 - 64$ **(2)**

**Common factors 2 and $p$** →
(a) $8p^3 + 6pqr = 2p(4p^2 + 3qr)$

**Check the signs** →
(b) $m^2 - 4m - 32 = (m + 4)(m - 8)$

**This is the difference of two squares** →
(c) $9x^2 - 64 = (3x + 8)(3x - 8)$

**4** A garden is in the shape of a rectangle with its length three times its width $x$m.

   **(a)** Write down an expression in terms of $x$ for the perimeter $(P)$ and then find the length of the garden if the perimeter is 160m. **(1)**

   **(b)** What does the width have to be to give an area of 1875m$^2$? **(3)**

**Perimeter is the distance around the sides of the rectangle** →
(a) $P = 2(x + 3x)$

$P = 2x + 6x$

$P = 8x$

$P = 8x = 160$

$\therefore x = \dfrac{160}{8} = 20\text{m}$

length of garden $= 3x = 60\text{m}$

(b) $A = x \times 3x = 3x^2 = 1875$

$x^2 = \dfrac{1875}{3} = 625$

$x = \sqrt{625}$

$\therefore$ width $= 25\text{m}$

# Exam practice questions

**1** Simplify: 🔲

(a) $a + 2b + 3a - 3b$

(b) $3y - x + y - 2y$

(c) $2(p - p^2) - (q^2 + p) + pq$ **(3)**

**2** (a) Make $a$ the subject of each formula: 🔲

    (i) $v^2 = u^2 + 2as$

    (ii) $x = \sqrt{\dfrac{a}{(a + b)}}$ **(5)**

(b) Solve the inequalities:

    (i) $5 > x - 3$

    (ii) $2(3y + 4) < 3$

    (iii) $4(x - 3) \leqslant 18 + 2x$ **(4)**

**3** A sequence begins 0, 2, 6, 12, ...

(a) Work out the next term.

(b) Work out the $n$th term of this sequence. **(3)**

**4** If $m = -1$, $n = 2$, $p = 3$, $q = \dfrac{1}{2}$ calculate the value of: 🔲

(a) $q(p - m)$

(b) $(m + n + p)^2$

(c) $\dfrac{mn + nq + p}{n}$ **(6)**

**5** Jonah buys 3 pens and 5 pencils for £3.25.

His sister buys 2 pens and 3 pencils for £2.05 in the same shop.

Form and solve two simultaneous equations to find the cost of a pen and a pencil. **(8)**

**6** Solve these equations:

(a) $3(3 + 2y) = y - 16$

(b) $x^2 + 4x - 12 = 0$

(c) $16a^2 - 36 = 0$

(d) $\dfrac{x + 12}{7} = \dfrac{x - 4}{3}$ **(8)**

**7** Match these sketches and equations of graphs: 🔲

(a) $y = -x^2$    (b) $y = \dfrac{10}{x}$    (c) $y = x^2 - 5x$    (d) $y = x^2 + 3$ **(4)**

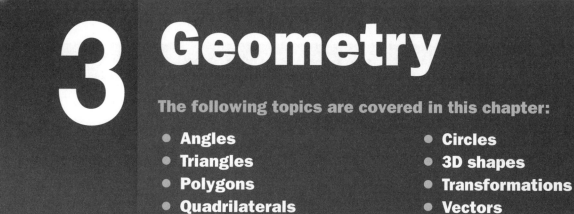

# 3 Geometry

**The following topics are covered in this chapter:**

- Angles
- Triangles
- Polygons
- Quadrilaterals
- Trigonometry
- Circles
- 3D shapes
- Transformations
- Vectors
- Constructions and loci

# 3.1 Angles

| LEARNING SUMMARY | **After studying this section, you should be able to understand:** <br> • facts about angles <br> • angle properties |
|---|---|

## Facts about angles

AQA UNITISED ✓
AQA LINEAR ✓
EDEXCEL A ✓
EDEXCEL B ✓
OCR A ✓
OCR B ✓
WJEC UNITISED ✓
WJEC LINEAR ✓
CCEA ✓

> **KEY POINT**
>
> An angle is the amount of turning measured in degrees.

| Angle | Fact | Diagram |
|---|---|---|
| Angles at a point | Add up to 360° <br> Called a **full revolution**. <br> $a + b + c + d = 360°$ | |
| Acute angles | Lie between 0° and 90° <br> $0° < a < 90°$ | |
| Obtuse angles | Lie between 90° and 180° <br> $90° < a < 180°$ | |
| Reflex angles | Lie between 180° and 360° <br> $180° < a < 360°$ | |
| Right angles | Equal 90° | |
| Complementary angles | Add up to 90° <br> $a + b = 90°$ | |
| Adjacent angles on a straight line | Add up to 180° <br> These angles are called **supplementary angles**. <br> $a + b + c = 180°$ | |

# Angle properties

- When two straight lines intersect, the opposite angles are equal. These are called vertically opposite angles.

  $a = c$     $b = d$

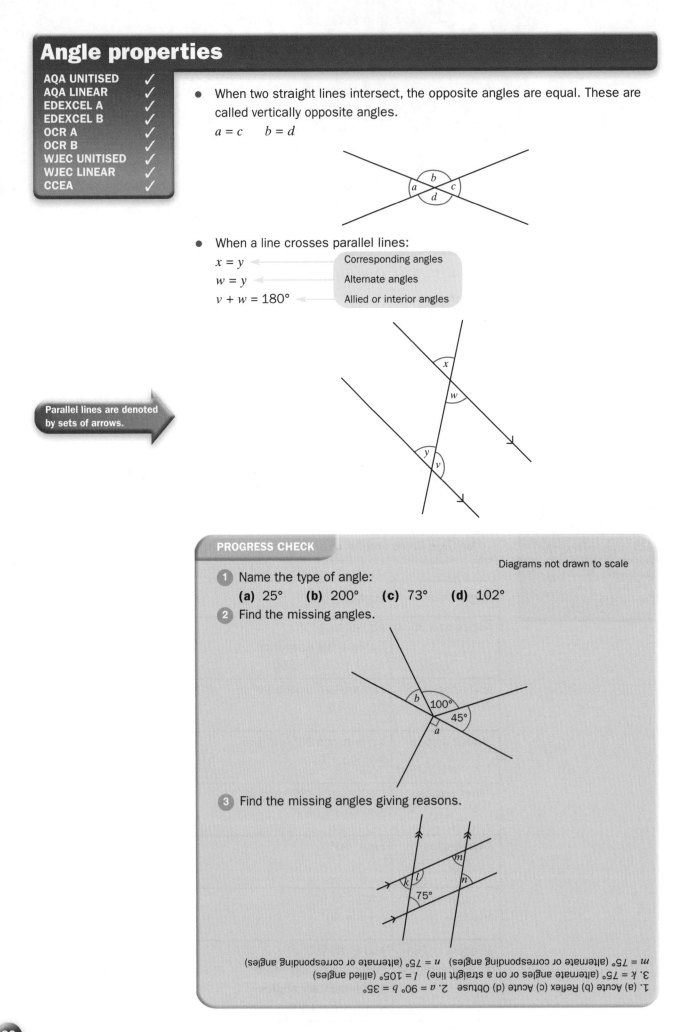

- When a line crosses parallel lines:

  $x = y$ ← Corresponding angles

  $w = y$ ← Alternate angles

  $v + w = 180°$ ← Allied or interior angles

> Parallel lines are denoted by sets of arrows.

## PROGRESS CHECK

Diagrams not drawn to scale

1. Name the type of angle:
   (a) 25°    (b) 200°    (c) 73°    (d) 102°
2. Find the missing angles.

3. Find the missing angles giving reasons.

---

1. (a) Acute (b) Reflex (c) Acute (d) Obtuse   2. $a = 90°$   $b = 35°$

3. $k = 75°$ (alternate angles or on a straight line)   $l = 105°$ (allied angles)

$m = 75°$ (alternate or corresponding angles)   $n = 75°$ (alternate or corresponding angles)

# 3.2 Triangles

**LEARNING SUMMARY**

After studying this section, you should be able to understand:

- notation used in triangles
- types of triangles
- Pythagoras' theorem
- congruency and similarity

## Notation used in triangles

AQA UNITISED ✓
AQA LINEAR ✓
EDEXCEL A ✓
EDEXCEL B ✓
OCR A ✓
OCR B ✓
WJEC UNITISED ✓
WJEC LINEAR ✓
CCEA ✓

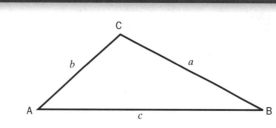

Angles are written as $\angle$ A or $\angle$ BAC ← This is sometimes seen as Â or BÂC

$\angle$ BAC is the angle where AB meets AC.

> You will often find that a small letter is written in the angle for convenience.

Sides are usually written using their endpoints AB, BC, CA.

Sometimes a small letter is used where '$a$' is the side opposite $\angle$ A.

Triangle ABC may be written as $\triangle$ ABC.

## Types of triangles

AQA UNITISED ✓
AQA LINEAR ✓
EDEXCEL A ✓
EDEXCEL B ✓
OCR A ✓
OCR B ✓
WJEC UNITISED ✓
WJEC LINEAR ✓
CCEA ✓

> **KEY POINT**
>
> A triangle is a polygon with three sides and three angles that add up to 180°

> Denote equal length sides by short cross lines.
>
> A triangle with three acute angles is called an acute-angled triangle.
>
> A triangle with an obtuse angle is called an obtuse-angled triangle.
>
> Remember, there can only be one obtuse angle because the angles in a triangle add up to 180°

| Triangle | Fact | Diagram |
|---|---|---|
| Scalene triangle | Three different sides and three different angles. $a + b + c = 180°$ | |
| Equilateral triangle | Three equal sides and three equal angles. As the three equal angles add up to 180°, they each equal 60° | |
| Isosceles triangle | Two equal sides. The angles opposite these sides are equal. | |
| Right-angled triangle | Has an angle of 90° The side opposite the right angle is called the **hypotenuse**. | Hypotenuse |

## Exterior angle of a triangle

> **KEY POINT**
>
> The exterior angle of a triangle is equal to the sum of the opposite two angles.

$x + y + z = 180°$

$a + z = 180°$

$\therefore a = x + y$

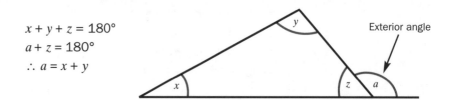

Exterior angle

# Pythagoras' theorem

> **KEY POINT**
>
> **Pythagoras' theorem** states that in a right-angled triangle, the square on the hypotenuse is equal to the sum of the squares on the other two sides.
>
> $BC^2 = AC^2 + AB^2$
>
> or $a^2 = b^2 + c^2$

Pythagoras' theorem can be used...

- to find the length of a side when you are given the length of the other two sides
- to prove that a triangle is right-angled
- to find the length of a line segment
- in 3D problems.

## Finding a missing side of a right-angled triangle

**Examples**

1. Find the hypotenuse:

   $BC^2 = AC^2 + AB^2$

   $\quad = 10^2 + 24^2$

   $\quad = 100 + 576 = 676$

   $\therefore BC = \sqrt{676} = 26cm$

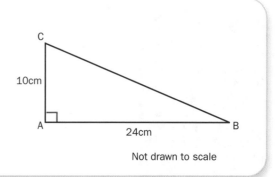

Not drawn to scale

> Do not forget to square root to find the answer.

**2.** Find the missing side:

$$AB^2 + BC^2 = AC^2$$
$$AB^2 = AC^2 - BC^2$$
$$= 12^2 - 8^2$$
$$= 144 - 64 = 80$$
$$\therefore AB = \sqrt{80} = 8.94\text{cm (2 d.p.)}$$

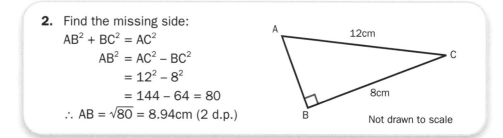

Not drawn to scale

## Proving that a triangle is right-angled

> Unless a triangle is drawn accurately, do not assume that an angle is a right angle.

> Take the longest side to be the hypotenuse.

> If the sides of the triangle are in the ratios
> 3 : 4 : 5      5 : 12 : 13
> 8 : 15 : 17    7 : 24 : 25
> you will find they are right-angled triangles.

> The symbol ≠ means 'not equal'.

**Example**                                    Diagrams not drawn to scale

Are these triangles right-angled?

**(a)**                                    **(b)**

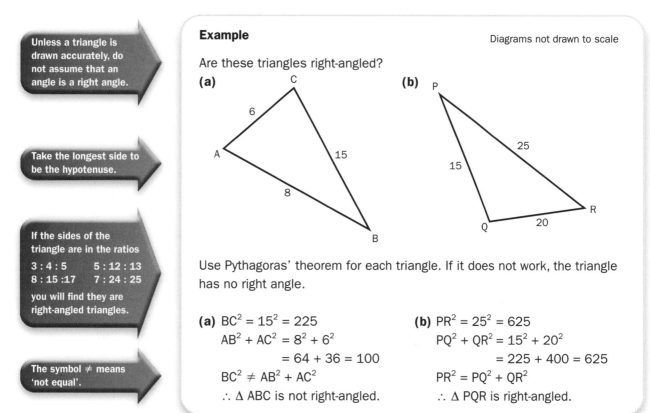

Use Pythagoras' theorem for each triangle. If it does not work, the triangle has no right angle.

**(a)** $BC^2 = 15^2 = 225$
     $AB^2 + AC^2 = 8^2 + 6^2$
                $= 64 + 36 = 100$
     $BC^2 \neq AB^2 + AC^2$
     $\therefore \Delta \text{ ABC is not right-angled.}$

**(b)** $PR^2 = 25^2 = 625$
     $PQ^2 + QR^2 = 15^2 + 20^2$
                $= 225 + 400 = 625$
     $PR^2 = PQ^2 + QR^2$
     $\therefore \Delta \text{ PQR is right-angled.}$

## Finding the length of a line segment

> **KEY POINT**
>
> A **line segment** is part of a line. This may be given by the coordinates of its ends.

You can use Pythagoras' theorem to find the length of a line segment and the coordinates of the midpoint of a line segment.

> **KEY POINT**
>
> The $x$-coordinate of the midpoint of a line segment is given by the average of the $x$-coordinates of the two endpoints.
>
> The $y$-coordinate of the midpoint of a line segment is given by the average of the $y$-coordinates of the two endpoints.
>
> The average is found by adding the coordinates together and dividing by two.

**Example**

(a) Find the length of the line segments AB and PQ.

(b) What are the coordinates of their midpoints ($M_a$ and $M_b$)?

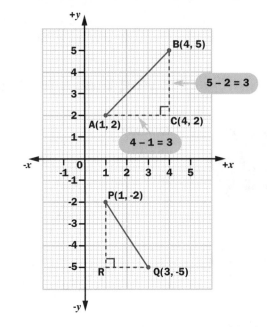

Form a right-angled triangle using the line segment as the hypotenuse.
Write out Pythagoras' theorem and then substitute in the values.

(a) $AB^2 = AC^2 + BC^2$            $PQ^2 = PR^2 + RQ^2$

     $AB^2 = 3^2 + 3^2 = 9 + 9$        $PQ^2 = (-3)^2 + 2^2 = 9 + 4$

           $= 18$                       $= 13$

       $\therefore AB = \sqrt{18} = 4.2$ (1 d.p.)     $PQ = \sqrt{13} = 3.6$ (1 d.p.)

(b) Endpoints of AB are A(1, 2) and B(4, 5)

     Endpoints of PQ are P(1, -2) and Q(3, -5)

     Midpoint of AB: $x = (1 + 4) \div 2 = 2.5$; $y = (2 + 5) \div 2 = 3.5$

     Midpoint of PQ: $x = (1 + 3) \div 2 = 2$; $y = (-2 + -5) \div 2 = -3.5$

     $M_a$ is the point (2.5, 3.5)

     $M_b$ is the point (2, -3.5)

## 3D problems

**Example**

Find the lengths of the diagonals BD and BE.     Diagrams not drawn to scale

**Draw out the triangles. This will show you exactly what you are looking for.**

**You may not need to square root the first stage if you just need the square in the second stage.**

$BD^2 = AB^2 + AD^2$            $BE^2 = BD^2 + DE^2$

       $= 7^2 + 4^2$                   $= 65 + 3^2$

       $= 49 + 16$               $= 65 + 9$

       $= 65$                    $= 74$

$\therefore BD = \sqrt{65} = 8.06$mm (3 s.f.)     $BE = \sqrt{74} = 8.6$mm (1 d.p)

# Congruency and similarity

## Congruent triangles

> **KEY POINT**
>
> Triangles that are exactly the same shape and size are said to be **congruent**. This also applies to other shapes.

> The side and angles must be in the same position in both triangles.

Triangles can be proved to be congruent if...

- two angles and a corresponding side are equal (AAS)

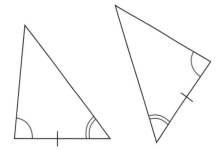

- two sides and the included angle are equal (SAS)

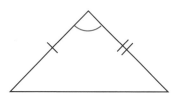
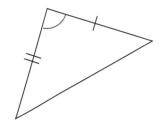

- three sides are equal (SSS)

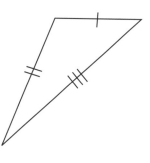

- the hypotenuse and one side are equal (in right-angled triangles) (RHS)

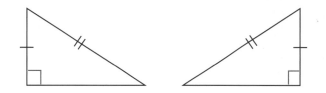

## Similar triangles

> **KEY POINT**
>
> Triangles that have the same angles, but different length sides, are said to be **similar**. The lengths of the sides are in the same ratio or proportion. This also applies to other shapes.

If two triangles are similar, one is an enlargement of the other. The lengths will enlarge by a given scale factor.

**Example**

Diagrams not drawn to scale

Find PQ, XZ, YZ.

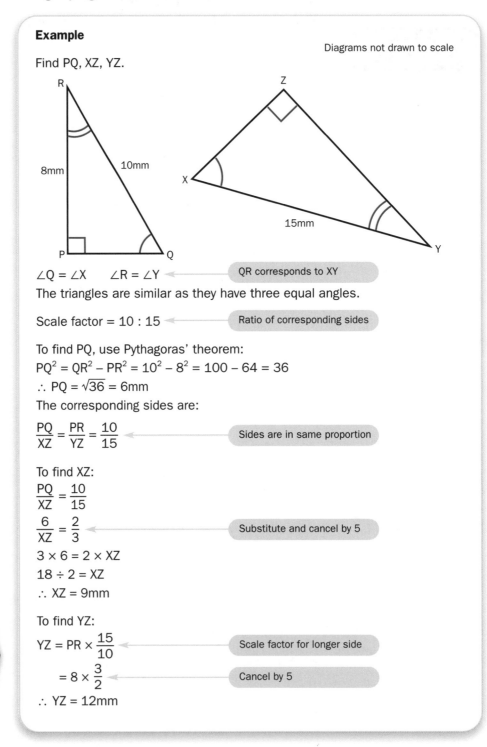

$\angle Q = \angle X \qquad \angle R = \angle Y$ ← QR corresponds to XY

The triangles are similar as they have three equal angles.

Scale factor = 10 : 15 ← Ratio of corresponding sides

To find PQ, use Pythagoras' theorem:
$PQ^2 = QR^2 - PR^2 = 10^2 - 8^2 = 100 - 64 = 36$
$\therefore PQ = \sqrt{36} = 6mm$

The corresponding sides are:

$\dfrac{PQ}{XZ} = \dfrac{PR}{YZ} = \dfrac{10}{15}$ ← Sides are in same proportion

To find XZ:
$\dfrac{PQ}{XZ} = \dfrac{10}{15}$

$\dfrac{6}{XZ} = \dfrac{2}{3}$ ← Substitute and cancel by 5

$3 \times 6 = 2 \times XZ$
$18 \div 2 = XZ$
$\therefore XZ = 9mm$

To find YZ:

$YZ = PR \times \dfrac{15}{10}$ ← Scale factor for longer side

$\quad = 8 \times \dfrac{3}{2}$ ← Cancel by 5

$\therefore YZ = 12mm$

In this example, you are finding a side on a larger triangle, so use $\frac{15}{10}$ as the scale factor.

If you were finding a side on a smaller triangle, you would use $\frac{10}{15}$ as the scale factor.

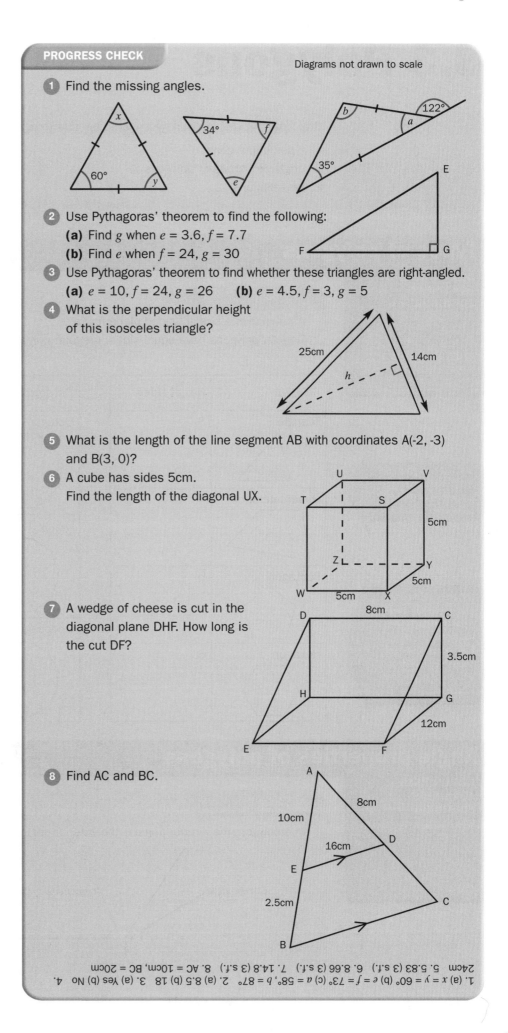

**PROGRESS CHECK**

Diagrams not drawn to scale

1  Find the missing angles.

2  Use Pythagoras' theorem to find the following:
   (a) Find $g$ when $e = 3.6$, $f = 7.7$
   (b) Find $e$ when $f = 24$, $g = 30$

3  Use Pythagoras' theorem to find whether these triangles are right-angled.
   (a) $e = 10$, $f = 24$, $g = 26$    (b) $e = 4.5$, $f = 3$, $g = 5$

4  What is the perpendicular height of this isosceles triangle?

5  What is the length of the line segment AB with coordinates A(-2, -3) and B(3, 0)?

6  A cube has sides 5cm.
   Find the length of the diagonal UX.

7  A wedge of cheese is cut in the diagonal plane DHF. How long is the cut DF?

8  Find AC and BC.

1. (a) $x = y = 60°$ (b) $e = f$, $b = 58°$, (c) $a = 73°$ 2. (a) 8.5 (b) 18 3. (a) Yes (b) No 4. 24cm 5. 5.83 (3 s.f.) 6. 8.66 (3 s.f.) 7. 14.8 (3 s.f.) 8. AC = 10cm, BC = 20cm

# 3.3 Polygons

**LEARNING SUMMARY**

After studying this section, you should be able to understand:

- types of polygons
- angle properties of polygons
- tessellations

## Types of polygons

| | |
|---|---|
| AQA UNITISED | ✓ |
| AQA LINEAR | ✓ |
| EDEXCEL A | ✓ |
| EDEXCEL B | ✓ |
| OCR A | ✓ |
| OCR B | ✓ |
| WJEC UNITISED | ✓ |
| WJEC LINEAR | ✓ |
| CCEA | ✓ |

**KEY POINT**

A **polygon** is a straight-sided plane figure.
Regular polygons have equal sides. Irregular polygons do not.

> See 3.2 Triangles and 3.4 Quadrilaterals for properties of these polygons.

> If AQA is your exam board, you will also need to recognise nonagons (9 sides). If Edexcel is your exam board, you will need to recognise heptagons (7 sides).

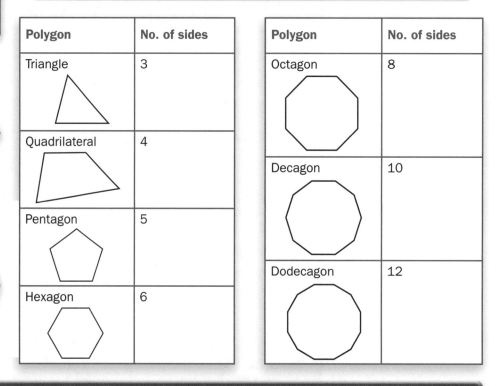

| Polygon | No. of sides | | Polygon | No. of sides |
|---|---|---|---|---|
| Triangle | 3 | | Octagon | 8 |
| Quadrilateral | 4 | | Decagon | 10 |
| Pentagon | 5 | | Dodecagon | 12 |
| Hexagon | 6 | | | |

## Angle properties of polygons

| | |
|---|---|
| AQA UNITISED | ✓ |
| AQA LINEAR | ✓ |
| EDEXCEL A | ✓ |
| EDEXCEL B | ✓ |
| OCR A | ✓ |
| OCR B | ✓ |
| WJEC UNITISED | ✓ |
| WJEC LINEAR | ✓ |
| CCEA | ✓ |

**KEY POINT**

In a polygon, the interior angle + the exterior angle = 180°

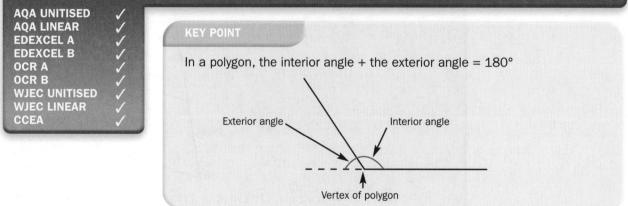

Exterior angle    Interior angle

Vertex of polygon

## Exterior angle of a polygon

> **KEY POINT**
>
> For all polygons, the sum of the exterior angles = 360°

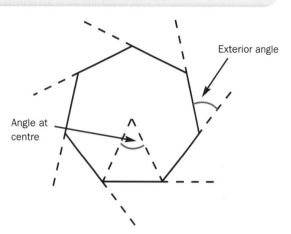

The angle at the centre of a regular polygon = exterior angle = 360° ÷ number of sides.

**Example**

Calculate the exterior angle of a regular pentagon.

A regular pentagon has five sides and five equal exterior angles.
Sum of exterior angles = 360°
Exterior angle = 360° ÷ 5 = 72°

## Interior angle of a polygon

To find the interior angle of a regular polygon, find the exterior angle first.

**Example**

Calculate the interior angle of a regular hexagon.

A regular hexagon has six sides and six equal exterior angles.
Exterior angle = 360° ÷ 6 = 60°
Interior angle = 180° − 60° = 120°

> Learn that the number of lines of symmetry of a regular polygon is equal to the number of sides.

To work out the sum of the interior angles of an irregular polygon, first choose one vertex (corner) of the polygon. Draw the diagonals from that vertex to the other vertices. This divides the polygon into triangles.

For example:
A seven-sided polygon (heptagon) has five triangles.

> Try working out the sum of the interior angles of different-sided polygons for yourself.

Each triangle has an angle sum of 180°, so sum of interior angles = 5 × 180° = 900°

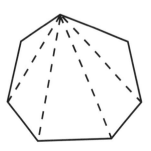

This table shows the sum of the interior angles for different irregular polygons:

| Number of sides | Triangles | Sum of interior angles |
|---|---|---|
| 3 | 1 | 1 × 180° = 180° |
| 4 | 2 | 2 × 180° = 360° |
| 5 | 3 | 3 × 180° = 540° |
| 6 | 4 | 4 × 180° = 720° |
| 8 | 6 | 6 × 180° = 1080° |
| 10 | 8 | 8 × 180° = 1440° |
| 12 | 10 | 10 × 180° = 1800° |

Notice that the number of triangles is 2 less than the number of sides of the polygon.

∴ sum of interior angles of a polygon = $(n - 2) \times 180°$

where $n$ is the number of sides of a polygon.

### Example

Calculate the sum of the interior angles of a nine-sided irregular polygon.

Sum of interior angles = $(9 - 2) \times 180°$ ← Substitute $n = 9$

= $7 \times 180° = 1260°$

# Tessellations

AQA UNITISED ✓
AQA LINEAR ✓
EDEXCEL A ✓
EDEXCEL B ✓
OCR A ✗
OCR B ✗
WJEC UNITISED ✓
WJEC LINEAR ✓
CCEA ✗

All triangles and quadrilaterals will tessellate.

**KEY POINT**

A **tessellation** is a pattern made from fitting together polygons (usually regular polygons) without leaving gaps.

Tessellations are used for tiling walls and floors.

The points where vertices meet must have a total angle of 360°.

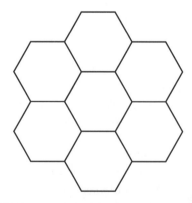

**PROGRESS CHECK**

1. What is the exterior angle of a regular...
   (a) octagon?   (b) dodecagon?
2. Calculate the sum of the interior angles of an irregular...
   (a) pentagon.   (b) hexagon.
3. Draw a tessellation using squares and octagons.

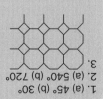

3.
2. (a) 540° (b) 720°
1. (a) 45° (b) 30°

# 3.4 Quadrilaterals

**LEARNING SUMMARY**

**After studying this section, you should be able to understand:**

- angles in a quadrilateral
- properties of quadrilaterals

## Angles in a quadrilateral

| | |
|---|---|
| AQA UNITISED | ✓ |
| AQA LINEAR | ✓ |
| EDEXCEL A | ✓ |
| EDEXCEL B | ✓ |
| OCR A | ✓ |
| OCR B | ✓ |
| WJEC UNITISED | ✓ |
| WJEC LINEAR | ✓ |
| CCEA | ✓ |

**KEY POINT**

A **quadrilateral** is a polygon with four sides and four angles that add up to 360°.

We can easily prove that the angle sum of a quadrilateral equals 360°.

Divide any quadrilateral into two parts by drawing a diagonal.

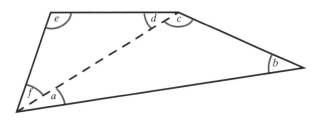

This gives two triangles each of which has an angle sum of 180°

$a + b + c = 180°$

$d + e + f = 180°$

∴ the angle sum of a quadrilateral = $2 \times 180° = 360°$

---

**Example**

Find ∠ CBD and ∠ CAD.

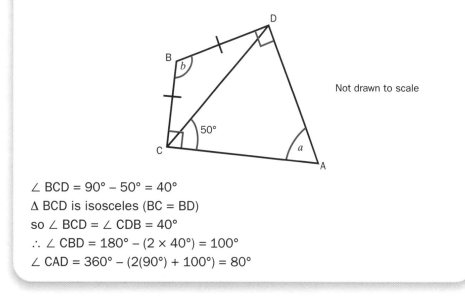

Not drawn to scale

∠ BCD = 90° − 50° = 40°

Δ BCD is isosceles (BC = BD)

so ∠ BCD = ∠ CDB = 40°

∴ ∠ CBD = 180° − (2 × 40°) = 100°

∠ CAD = 360° − (2(90°) + 100°) = 80°

---

# Properties of quadrilaterals

AQA UNITISED ✓
AQA LINEAR ✓
EDEXCEL A ✓
EDEXCEL B ✓
OCR A ✓
OCR B ✓
WJEC UNITISED ✓
WJEC LINEAR ✓
CCEA ✓

It is important to learn how to recognise different quadrilaterals.
This table shows you the properties of different quadrilaterals.

| Quadrilateral | Sides | Angles | Diagonals | Symmetry |
|---|---|---|---|---|
| Square | Four equal and parallel sides | All angles are 90° | Bisect each other at right angles | Four lines of symmetry. Order of rotation 4 |
| Rectangle | Two pairs of equal and parallel sides | All angles are 90° | Bisect each other | Two lines of symmetry. Order of rotation 2 |
| Parallelogram | Two pairs of equal and parallel sides | Opposite angles are equal | Bisect each other | No lines of symmetry. Order of rotation 2 |
| Rhombus | Four equal sides. Two pairs of parallel sides | Opposite angles are equal | Bisect each other at right angles | Two lines of symmetry. Order of rotation 2 |
| Kite | Two pairs of adjacent sides equal | One pair of opposite angles are equal | Cut at right angles; shorter diagonal bisected | One line of symmetry. No rotational symmetry |
| Arrowhead | Two pairs of adjacent sides are equal | One pair of equal angles | One diagonal | One line of symmetry. No rotational symmetry |
| Trapezium | One pair of parallel sides | No equal angles | Diagonals different | No lines of symmetry. No rotational symmetry |
| Isosceles trapezium | One pair of parallel sides. 2nd pair equal sides | Two pairs of equal angles | Two equal diagonals | One line of symmetry. No rotational symmetry |

## PROGRESS CHECK

Find $a$, $b$ and $c$ in each quadrilateral.

1. 72° 2. 130°, 63° 3. 56°

1. $a = 72°$ $b = 108°$ $c = 72°$   2. $a = 77°$ $b = 103°$ $c = 103°$   3. $a = 34°$ $b = 34°$ $c = 112°$

# 3.5 Trigonometry

After studying this section, you should be able to understand:

- trigonometric ratios
- finding trigonometric ratios of any angle
- area of a triangle using trigonometric formula
- sine and cosine rules

## Trigonometric ratios

| | |
|---|---|
| AQA UNITISED | ✓ |
| AQA LINEAR | ✓ |
| EDEXCEL A | ✓ |
| EDEXCEL B | ✓ |
| OCR A | ✓ |
| OCR B | ✓ |
| WJEC UNITISED | ✓ |
| WJEC LINEAR | ✓ |
| CCEA | ✓ |

**Trigonometry** shows the relationship between sides and angles in a right-angled triangle. Trigonometry can be used to find the length of sides and sizes of angles in right-angled triangles.

> If only sides are mentioned in the question, use Pythagoras' theorem; if angles are mentioned in the question as well, use trigonometry.

**KEY POINT**

Learn these ratios:

$$\sin \theta = \frac{\text{Opposite}}{\text{Hypotenuse}}$$

$$\cos \theta = \frac{\text{Adjacent}}{\text{Hypotenuse}}$$

$$\tan \theta = \frac{\text{Opposite}}{\text{Adjacent}}$$

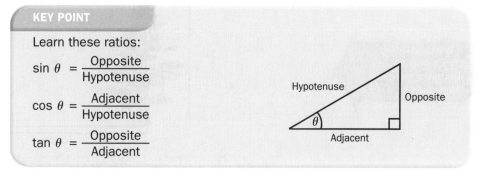

There are two ways to use these ratios.

### Finding a side

> Check your calculator to find the sin, cos and tan keys.

**Example**

Find AB.

$$\frac{AB}{8} = \sin 26°$$ ← Use $\sin = \frac{\text{Opposite}}{\text{Hypotenuse}}$

$$AB = 8 \times \sin 26°$$

$$\quad = 3.5069$$

$$AB = 3.51\text{cm (3 s.f.)}$$

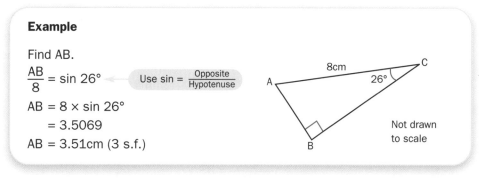

Not drawn to scale

### Finding an angle

> You could cut out the middle step but marks are awarded for giving the correct ratio.

**Example**

Find ∠ BAC.

$$\cos A = \frac{7.3}{18.2}$$ ← Use $\cos = \frac{\text{Adjacent}}{\text{Hypotenuse}}$

$$\quad = 0.4011$$ ← Use the $\cos^{-1}$ key on your calculator

$$\angle BAC = 66.4°$$ ← You should usually give an angle to 1 d.p.

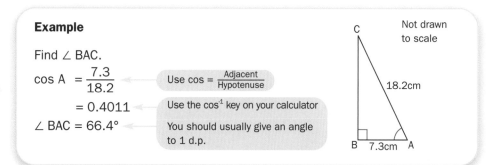

Not drawn to scale

# Angles of elevation and depression

> **KEY POINT**
>
>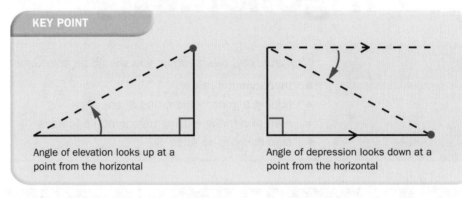
>
> Angle of elevation looks up at a point from the horizontal
>
> Angle of depression looks down at a point from the horizontal

**Example**                                    Diagrams not drawn to scale

1. What is the angle of elevation of the top of a tree, 21m high, from a point 25m from the foot of the tree?

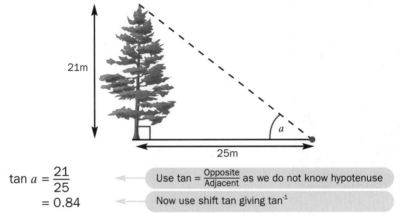

*It is a good idea to draw a diagram illustrating the given information.*

$$\tan a = \frac{21}{25}$$ ← Use $\tan = \frac{\text{Opposite}}{\text{Adjacent}}$ as we do not know hypotenuse

$$= 0.84$$ ← Now use shift tan giving $\tan^{-1}$

Angle of elevation ($a$) = 40° (to nearest degree)

2. A man standing on the top of a cliff 40m high looks at a yacht out at sea. The angle of depression is 22°. How far away from the base of the cliff is the boat? (Assume the cliff meets the ground at 90° and the horizontal and ground are parallel for the purposes of the calculation.)

RS is the distance of the boat from the cliff.

∠ RST = 22° (alternate angles)

$$\frac{40}{RS} = \tan 22°$$ ← Use $\tan = \frac{\text{Opposite}}{\text{Adjacent}}$

$$\frac{40}{\tan 22°} = RS$$ ← Rearrange formula

Distance of boat (RS) = 99m (to nearest whole metre)

# Finding trigonometric ratios of any angle

| | |
|---|---|
| AQA UNITISED | ✓ |
| AQA LINEAR | ✓ |
| EDEXCEL A | ✗ |
| EDEXCEL B | ✗ |
| OCR A | ✗ |
| OCR B | ✗ |
| WJEC UNITISED | ✓ |
| WJEC LINEAR | ✓ |
| CCEA | ✗ |

This diagram shows a circle with radius, $r$, with centre at the origin. The point $P$, which has coordinates $(x, y)$, can be anywhere on the circumference.

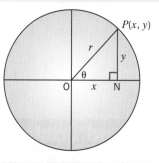

**KEY POINT**

Using the right-angled triangle NOP:

$$\sin \theta = \frac{NP}{OP}, = \frac{y}{r} \qquad \cos \theta = \frac{ON}{OP} = \frac{x}{r} \qquad \tan \theta = \frac{NP}{ON} = \frac{y}{x}$$

The circle can be divided up into quadrants like this:

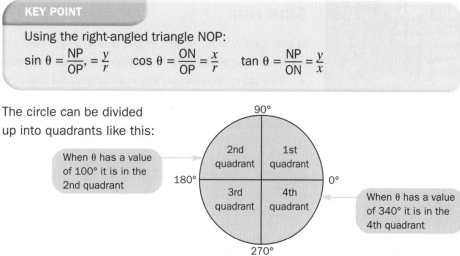

When θ has a value of 100° it is in the 2nd quadrant

When θ has a value of 340° it is in the 4th quadrant

As $P$ moves round the circumference of the circle, so $x$ and $y$ can be negative or positive depending on the value of θ. For example, when θ is in the 2nd quadrant, $x$ is negative, but $y$ is positive.

This means that in the 2nd quadrant, $\sin \theta = \frac{y}{r}$ is positive, but $\cos \theta = \frac{x}{r}$ is negative (by convention we always assume that the radius is positive).

This table shows whether the angles are positive or negative in each quadrant:

| Quadrant | 1st | 2nd | 3rd | 4th |
|---|---|---|---|---|
| $\sin \theta$ | positive | positive | negative | negative |
| $\cos \theta$ | positive | negative | negative | positive |
| $\tan \theta$ | positive | negative | positive | negative |

# Area of a triangle using trigonometric formula

| | |
|---|---|
| AQA UNITISED | ✓ |
| AQA LINEAR | ✓ |
| EDEXCEL A | ✓ |
| EDEXCEL B | ✓ |
| OCR A | ✓ |
| OCR B | ✓ |
| WJEC UNITISED | ✓ |
| WJEC LINEAR | ✓ |
| CCEA | ✓ |

**KEY POINT**

The area of any triangle with two given sides and the angle between them can be found by using the formula:

Area $= \frac{1}{2} ab \sin C$

or $= \frac{1}{2} bc \sin A$

or $= \frac{1}{2} ac \sin B$

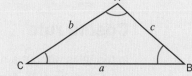

The formula you use depends on the information given.

**Example**

Find the area of $\triangle$ ABC

Area $= \frac{1}{2}bc \sin A$

$= \frac{1}{2} \times 7.2 \times 8 \times \sin 105°$

$= 27.8\text{mm}^2$ (3 s.f.)

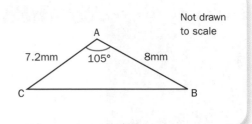

Not drawn to scale

# Sine and cosine rules

## Sine rule

In triangles without a right angle, use the sine rule to find:

- a missing side if you know two angles and an opposite side
- a missing angle if you know an angle and two sides.

> Remember, for right-angled triangles, use Pythagoras' theorem to find a missing side and trigonometric ratios to find a missing side or angle.

**KEY POINT**

The sine rule is:

$$\frac{a}{\sin A} = \frac{b}{\sin B} = \frac{c}{\sin C}$$

or

$$\frac{\sin A}{a} = \frac{\sin B}{b} = \frac{\sin C}{c}$$

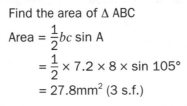

The triangle might use different letters such as PQR or CDE. Adjust the formula accordingly.

**Examples**

Diagrams not drawn to scale

1. Find AB.

   AB is side $c$ opposite $\angle$ C

   $$\frac{c}{\sin C} = \frac{a}{\sin A}$$

   Use the appropriate part of the sine rule

   $$\frac{c}{\sin 36°} = \frac{2.5}{\sin 77°}$$

   $$c = \frac{2.5 \times \sin 36°}{\sin 77°}$$

   $\therefore$ AB = 1.51m (3 s.f.)

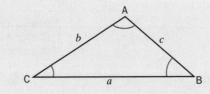

2. Find $\angle$ QPR.

   $$\frac{\sin P}{p} = \frac{\sin R}{r}$$

   $$\frac{\sin P}{16} = \frac{\sin 40°}{14}$$

   $$\sin P = \frac{16 \times \sin 40°}{14}$$

   $$= 0.734614411$$

   $\angle$ QPR = 47.3°

   Use the sin$^{-1}$ key on your calculator

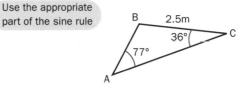

> Remember angles are given to 1 d.p. where necessary.

## Cosine rule

In triangles without a right angle, use the cosine rule to find:

- a missing side if you know two sides and the included angle
- a missing angle if you know three sides.

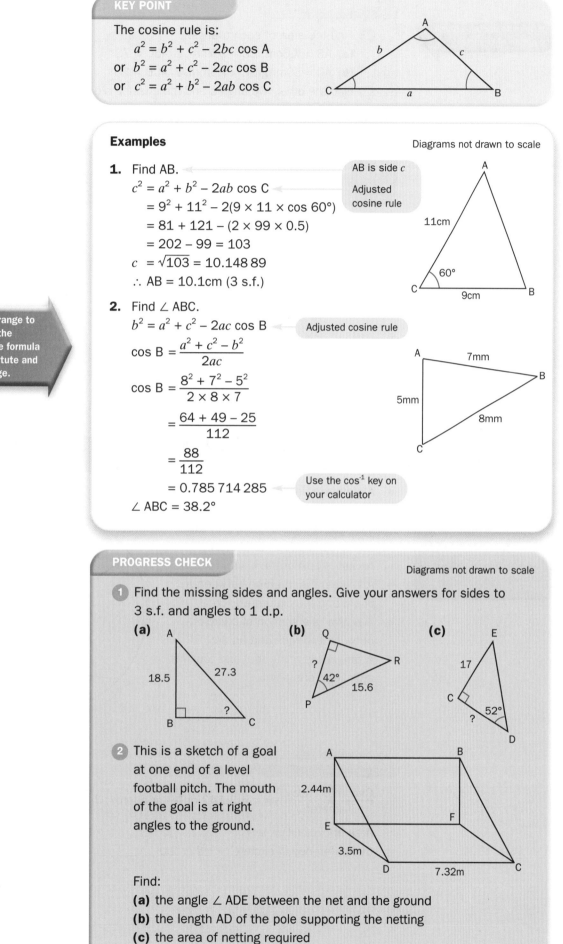

### KEY POINT

The cosine rule is:

$$a^2 = b^2 + c^2 - 2bc \cos A$$
or $$b^2 = a^2 + c^2 - 2ac \cos B$$
or $$c^2 = a^2 + b^2 - 2ab \cos C$$

### Examples

Diagrams not drawn to scale

1. Find AB.

   AB is side $c$

   $c^2 = a^2 + b^2 - 2ab \cos C$ — Adjusted cosine rule

   $= 9^2 + 11^2 - 2(9 \times 11 \times \cos 60°)$
   $= 81 + 121 - (2 \times 99 \times 0.5)$
   $= 202 - 99 = 103$
   $c = \sqrt{103} = 10.148\,89$
   $\therefore$ AB = 10.1cm (3 s.f.)

2. Find $\angle$ ABC.

   $b^2 = a^2 + c^2 - 2ac \cos B$ — Adjusted cosine rule

   $\cos B = \dfrac{a^2 + c^2 - b^2}{2ac}$

   $\cos B = \dfrac{8^2 + 7^2 - 5^2}{2 \times 8 \times 7}$

   $= \dfrac{64 + 49 - 25}{112}$

   $= \dfrac{88}{112}$

   $= 0.785\,714\,285$ — Use the cos⁻¹ key on your calculator

   $\angle$ ABC = 38.2°

You can rearrange to make cos B the subject of the formula or just substitute and then rearrange.

### PROGRESS CHECK

Diagrams not drawn to scale

1. Find the missing sides and angles. Give your answers for sides to 3 s.f. and angles to 1 d.p.

   (a)
   18.5, 27.3, ?

   (b)
   ?, 42°, 15.6

   (c)
   17, 52°, ?

2. This is a sketch of a goal at one end of a level football pitch. The mouth of the goal is at right angles to the ground.

   2.44m, 3.5m, 7.32m

   Find:

   (a) the angle $\angle$ ADE between the net and the ground

   (b) the length AD of the pole supporting the netting

   (c) the area of netting required

It helps to sketch a rough diagram of triangles.

# 3.6 Circles

## Parts of a circle

| | |
|---|---|
| AQA UNITISED | ✓ |
| AQA LINEAR | ✓ |
| EDEXCEL A | ✓ |
| EDEXCEL B | ✓ |
| OCR A | ✓ |
| OCR B | ✓ |
| WJEC UNITISED | ✓ |
| WJEC LINEAR | ✓ |
| CCEA | ✓ |

- A **diameter** divides a circle into two semi-circles. If two diameters intersect at right angles, they divide the circle into four quadrants.
- An **arc** is part of the circumference.
- A **segment** is the area between a chord and the circumference.
- A **sector** is a section of a circle between two radii and an arc.
- A **tangent** is a line that just touches a circle at one point.

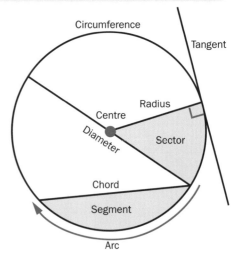

**KEY POINT**

$$\frac{\text{Circumference}(C)}{\text{Diameter}(d)} = \pi = 3.142 \text{ or } \frac{22}{7} \text{ or approx. } 3$$

Area of circle$(A) = \pi r^2$

Circumference of circle$(C) = \pi d$ or $2\pi r$

# Angles in a circle

## Angle in a semi-circle

> **KEY POINT**
>
> The angle in a semi-circle is always a right angle.

This can be proved as follows:

The angle in a semi-circle is subtended by a diameter at the circumference.

OA = OB = OC  ← All radii of a circle

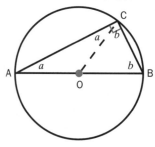

This means that △ OAC and △ OBC are isosceles.

In △ ABC $a + a + b + b = 180°$

giving $2(a + b) = 180°$

∴ ∠ ACB = $a + b = 90°$

## Angle at the centre of a circle

> **KEY POINT**
>
> The angle at the centre of a circle is always twice the size of the angle at the circumference subtended by the same arc.

This can be proved as follows:

OA = OB = OC  ← All radii of a circle

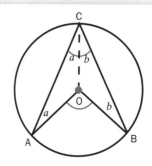

This means that △ OAC and △ OBC are isosceles. Their base angles are equal.

∠ AOC = $180° - 2a$

∠ BOC = $180° - 2b$

We know that ∠ AOB + ∠ AOC + ∠ BOC = 360°

giving ∠ AOB + $(180° - 2a)$ + $(180° - 2b) = 360°$

∴ ∠ AOB = $360° - 180° + 2a - 180° + 2b$

∴ ∠ AOB = $2(a + b) = 2 \times$ ∠ ACB

## Angles in the same segment

> **KEY POINT**
>
> Angles subtended by the same arc, at the circumference, in the same segment are always equal.

We know that the angle at the centre of a circle is twice the size of the angle at the circumference subtended by the same arc. This applies to any angle at the circumference subtended by this arc. In this diagram, the angles marked $a$ are equal.

## Angles in a cyclic quadrilateral

A **cyclic quadrilateral** is a quadrilateral drawn inside a circle so that all four vertices are on the circumference of the circle.

The opposite angles of a cyclic quadrilateral add up to 180°.
The exterior angle of a cyclic quadrilateral equals the opposite interior angle.

This can be proved as follows:

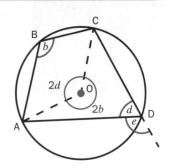

If we mark the centre of the circle (O) and sketch
in the radii OA and OC, we know that the angle at
the centre is twice the angle at the
circumference. We also know that
$2b + 2d = 360°$ or $b + d = 180°$
If we extend CD to form the exterior angle ($e$)
we know that $d + e = 180°$
∴ $b = e$

## Chord theorem

| | |
|---|---|
| AQA UNITISED | ✓ |
| AQA LINEAR | ✓ |
| EDEXCEL A | ✓ |
| EDEXCEL B | ✓ |
| OCR A | ✓ |
| OCR B | ✓ |
| WJEC UNITISED | ✓ |
| WJEC LINEAR | ✓ |
| CCEA | ✓ |

A line from the centre of a circle to the midpoint of a chord is perpendicular
to the chord.

This can be proved as follows:

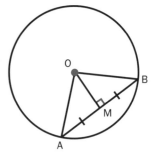

A perpendicular line is drawn from the centre (O)
to the midpoint (M) of a chord AB.

In Δ OAM and Δ OBM:
AM = MB
OA = OB ← Radii of circle
OM is the same for both triangles
∴ the triangles are congruent (SSS) so
∠ OMA = ∠ OMB = 180° ÷ 2 = 90°

## Tangents and angles

| | |
|---|---|
| AQA UNITISED | ✓ |
| AQA LINEAR | ✓ |
| EDEXCEL A | ✓ |
| EDEXCEL B | ✓ |
| OCR A | ✓ |
| OCR B | ✓ |
| WJEC UNITISED | ✓ |
| WJEC LINEAR | ✓ |
| CCEA | ✓ |

A tangent to a circle is perpendicular to a radius at the point.

Proving this is just an extension of the chord theorem. Draw a chord, then parallel
chords, until you get a tangent.

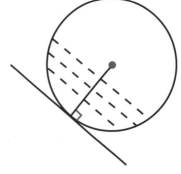

Two tangents drawn to a circle from the same point are equal.
The line drawn from this point, through the centre of the circle, bisects the angle between the tangents. This line bisects the chord joining the points of contact, at right angles.

AB = AC

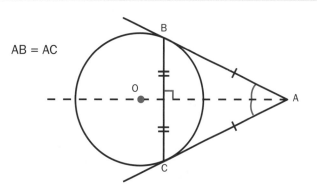

## Alternate segment theorem

The angle between a chord and a tangent at the point where it touches the circle equals any angle subtended by that chord in the alternate segment.

This can be proved as follows:

If AC is a diameter $\angle ABC = 90°$

↑

Angle in semi-circle

$\angle OAB + \angle ABC + \angle OCB = 180°$

↑

Angle sum of a triangle

so $\angle OAB + \angle OCB = 90°$
$\angle OCT = \angle OCB + \angle BCT = 90°$

↑

Tangent meeting radius

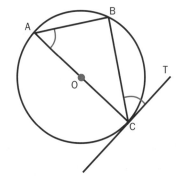

this means that $\angle BCT = \angle OAB$

From the angles in the same segment theorem, this will be true for any position of A.

Diagrams not drawn to scale

1. Find the missing angles in each of these diagrams.
   (a)                                    (b)

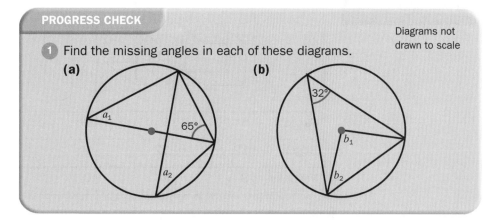

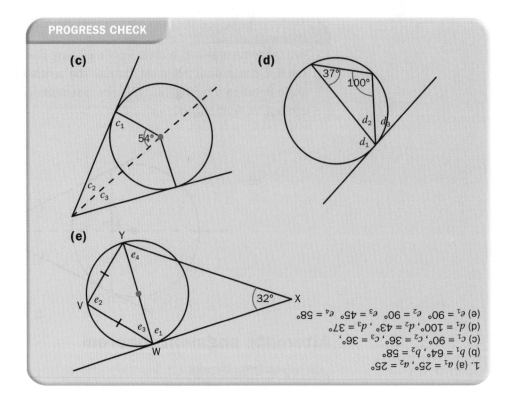

1. (a) $a_1 = 25°$, $a_2 = 25°$
(b) $b_1 = 64°$, $b_2 = 58°$
(c) $c_1 = 90°$, $c_2 = 36°$, $c_3 = 36°$,
(d) $d_1 = 100°$, $d_2 = 43°$, $d_3 = 37°$
(e) $e_1 = 90°$, $e_2 = 90°$, $e_3 = 45°$, $e_4 = 58°$

# 3.7 3D shapes

| LEARNING SUMMARY | After studying this section, you should be able to understand: |
|---|---|
| | • 3D shapes                 • isometric drawing |
| | • nets                      • plans and elevations |
| | • 3D coordinates |

## 3D shapes

| | |
|---|---|
| AQA UNITISED ✓ | |
| AQA LINEAR ✓ | Plane shapes have two dimensions (2D). |
| EDEXCEL A ✓ | Solid shapes have three dimensions (3D). |
| EDEXCEL B ✓ | |
| OCR A ✓ | For example, a square is a plane shape, but a cube is a solid shape. |
| OCR B ✓ | |
| WJEC UNITISED ✓ | |
| WJEC LINEAR ✓ | |
| CCEA ✓ | |

| 3D Shape | Cube | Cuboid | Cylinder | Sphere | Cone |
|---|---|---|---|---|---|
| | | | | (A hemisphere is $\frac{1}{2}$ a sphere) | |
| Faces | 6 | 6 | 3 | 1 | 2 |
| Edges | 12 | 12 | 2 | 0 | 1 |
| Vertices | 8 | 8 | 0 | 0 | 1 |

There are two more types of 3D shapes:

- **Prisms** take their name from the polygon shape of their uniform cross-section. For example, this diagram is a triangular prism.

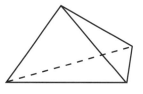

- **Pyramids** take their name from the polygon shape of their base. The perpendicular height is measured from the vertex to the base. A pyramid with four triangular faces is called a **tetrahedron**.

# Nets

| | |
|---|---|
| AQA UNITISED | ✓ |
| AQA LINEAR | ✓ |
| EDEXCEL A | ✓ |
| EDEXCEL B | ✓ |
| OCR A | ✓ |
| OCR B | ✓ |
| WJEC UNITISED | ✓ |
| WJEC LINEAR | ✓ |
| CCEA | ✗ |

A 3D solid shape can be constructed from a **net**. The net is the solid shape opened flat.

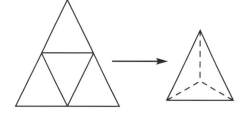

# 3D coordinates

| | |
|---|---|
| AQA UNITISED | ✓ |
| AQA LINEAR | ✓ |
| EDEXCEL A | ✓ |
| EDEXCEL B | ✓ |
| OCR A | ✓ |
| OCR B | ✓ |
| WJEC UNITISED | ✗ |
| WJEC LINEAR | ✗ |
| CCEA | ✗ |

A solid figure can be drawn using **3D coordinates**.

> **KEY POINT**
>
> Draw a 2D figure using 2D coordinates on two axes $x$ and $y$. A point has the coordinates $(x, y)$.
>
> Draw a 3D figure using 3D coordinates on three axes $x$, $y$, $z$. A point has the coordinates $(x, y, z)$.

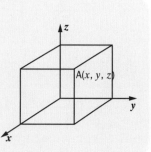

Sometimes, you may see the $x$, $y$ and $z$ axes in different positions.

**Example**

Shape ABCDEFGH is a cube.

**(a)** How long is a side?

A side is 4 units. ← Read off the axes

**(b)** What are the coordinates of B, E, F and H?

B (4, 4, 0) ← Read coordinate off each axis in turn

E (4, 0, 4)

F (4, 4, 4)

H (0, 0, 4)

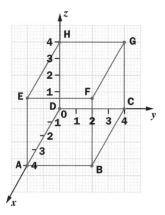

# Isometric drawing

| | |
|---|---|
| AQA UNITISED | ✓ |
| AQA LINEAR | ✓ |
| EDEXCEL A | ✓ |
| EDEXCEL B | ✓ |
| OCR A | ✓ |
| OCR B | ✓ |
| WJEC UNITISED | ✓ |
| WJEC LINEAR | ✓ |
| CCEA | X |

Isometric paper has dots or triangles. This enables you to draw solid figures.

Here is a drawing of a cuboid on three different types of paper:

**Graph paper**

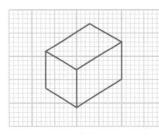

**Triangular isometric paper**

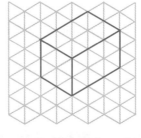

**Dotted isometric paper**

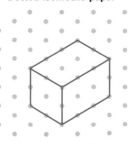

# Plans and elevations

| | |
|---|---|
| AQA UNITISED | ✓ |
| AQA LINEAR | ✓ |
| EDEXCEL A | ✓ |
| EDEXCEL B | ✓ |
| OCR A | ✓ |
| OCR B | ✓ |
| WJEC UNITISED | ✓ |
| WJEC LINEAR | ✓ |
| CCEA | X |

> **KEY POINT**
>
> Viewing a solid figure from above is a **plan**.
> Viewing a solid figure from the front is a **front elevation**.
> Viewing a solid figure from the side is a **side elevation**.

> In this example it does not matter which side is viewed, but sometimes it does. Read the question carefully.

These types of drawings are used in designing buildings. For example:

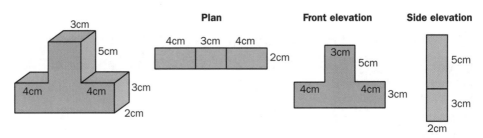

> **PROGRESS CHECK**
>
> 1. Fill the gaps in this table.
>
> | Faces | Edges | Vertices | Shape |
> |---|---|---|---|
> | 5 | 9 | 6 | |
> | 6 | | 8 | Cube |
> | 4 | 6 | | Tetrahedron |
> | | 1 | 0 | Hemisphere |
>
> 2. Draw the plan, front and side elevations for this shape made from unit cubes.
>
>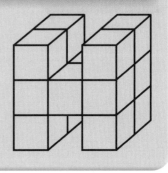

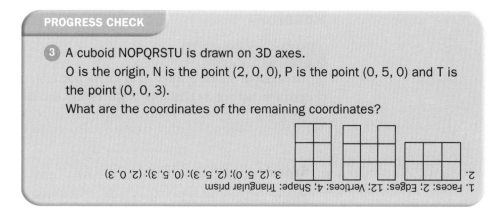
# 3.8 Transformations

**After studying this section, you should be able to understand:**

- types of transformations

## Types of transformations

A **transformation** moves a shape. It may or may not change size.

### Reflection

**KEY POINT**

**Reflection** of a shape about a line produces the mirror image of the shape. The line is called the mirror line. A reflection is described by giving the line of symmetry, sometimes as an equation.

**Example**

Reflect shape ABCDEF in the $x$-axis and in the $y$-axis.

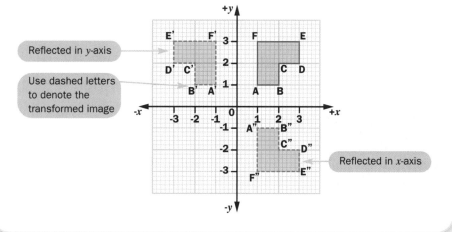

Note that the reflected shape remains the same distance from the line of symmetry as was the original shape. The reflected shape and the original shape are congruent, i.e. the same shape and size.

## Rotation

> **KEY POINT**
>
> **Rotation** of a shape about a centre, or point, of rotation changes its position but not its size. The angle, centre and direction of rotation must be given.

### Example

Rotate shape A, through a right angle, anticlockwise about the origin and 180° clockwise, also about the origin.

Draw a line from the centre of rotation to each vertex of the shape.
Measure the given angle in the given direction from the line for each vertex.
Connect the rotated vertices to give the rotated shape.

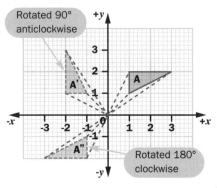

> Note that the rotated shape remains the same size as the original. The rotated shape and the original shape are congruent.

## Finding the centre of rotation

If the rotated shape is given and you are asked to find the centre of rotation, follow these steps:

1. Join a corresponding pair of vertices (original and rotated).
2. Construct the perpendicular bisector of this joining line (see page 114).
3. Repeat for another pair of vertices.
4. The two perpendicular bisectors cross at the centre of rotation.

## Translation

> **KEY POINT**
>
> **Translation** is when every point of a line or shape moves the same distance in the same direction. The movement is described by a **column vector**.

### Example

Translate △ ABC by the column vector $\binom{4}{3}$

This means move the shape 4 units in a positive horizontal direction and 3 units in a positive vertical direction, i.e. move each vertex 4 units horizontally to the right and 3 units upwards.
Label these A', B', C'.
Join A' B' C'.
A'B'C' is the translated triangle.

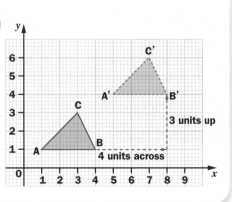

> Note that the translated shape remains the same size as the original. The translated shape and the original shape are congruent.

## Enlargement

> **KEY POINT**
>
> **Enlargement** of a shape is when every length is multiplied by the same scale factor.
> - If a scale factor > 1 the shape enlarges.
> - If a scale factor = 1 the shape stays the same size.
> - If 0 < scale factor < 1 the shape reduces in size.
> - If a scale factor < 0 the enlarged shape is upside down on the other side of the centre of enlargement.

### Example

Using centre of enlargement O, enlarge △ ABC by scale factor

**(a)** 3 **(b)** $\frac{1}{2}$ **(c)** -2

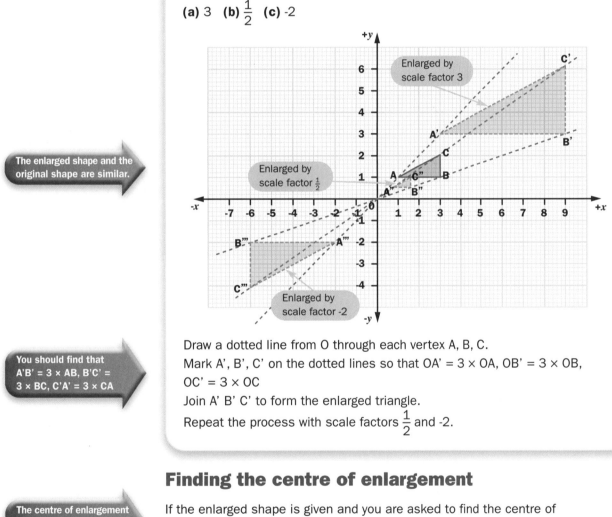

> The enlarged shape and the original shape are similar.

> You should find that A'B' = 3 × AB, B'C' = 3 × BC, C'A' = 3 × CA

Draw a dotted line from O through each vertex A, B, C.
Mark A', B', C' on the dotted lines so that OA' = 3 × OA, OB' = 3 × OB, OC' = 3 × OC
Join A' B' C' to form the enlarged triangle.
Repeat the process with scale factors $\frac{1}{2}$ and -2.

## Finding the centre of enlargement

> The centre of enlargement can be any point, not just the origin.

If the enlarged shape is given and you are asked to find the centre of enlargement and scale factor, follow these steps:

1 Draw dotted lines through two pairs of corresponding vertices.
2 The centre of enlargement is where these lines cross.

Scale factor = $\dfrac{\text{length of one side of enlarged shape}}{\text{length of corresponding side of original shape}}$

# Combining transformations

Two or more transformations may be performed on a shape. The combined effect may be the same as a single transformation.

1. Plot and join these points to form a shape:
   A (0, 0); B (2, 1); C (3, 3); D (1, 2); A (0, 0)
   **(a)** Reflect the shape in the $x$-axis and label it RE.
   **(b)** Rotate shape RE clockwise through 180° about the origin. Label it RO.
   **(c)** Translate shape RO by vector $\begin{pmatrix} 1 \\ -3 \end{pmatrix}$
   **(d)** What transformation would move the original shape to RO?

2. Plot and join these points to form a shape:
   P (4, 2), Q (5, 2), R (5, 3), S (6, 3), T (6, 5), U (5, 5), V (5, 6), W (4, 6), P (4, 2). Using centre of enlargement (-2, -2) transform the shape by scale factor $\frac{1}{2}$.

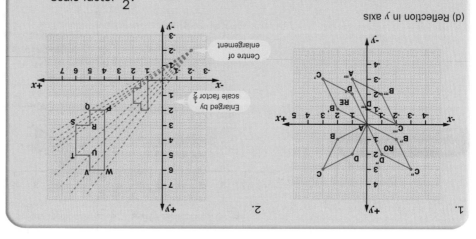

(d) Reflection in $y$ axis

# 3.9 Vectors

**After studying this section, you should be able to understand:**

- vector notation
- addition and subtraction of vectors
- real-life problems using vectors

## Vector notation

**KEY POINT**

A **vector** gives magnitude (size) and direction.

When a shape is translated, a column vector such as $\binom{2}{3}$ is used to describe the translation.

Vectors are also written using $\overrightarrow{AB}$ meaning that the vector starts at A and ends at B. They may also be written in the form **a** or a͜

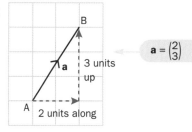

$$a = \binom{2}{3}$$

The **inverse of a vector** has the same size as the vector but the opposite direction.
The inverse of $\overrightarrow{AB}$ is $\overrightarrow{BA}$ or $-\overrightarrow{AB}$ or **-a**.

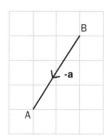

**Zero vector** has zero length and direction.
**Unit vector** has magnitude 1.

### Scalars

**KEY POINT**

A **scalar** is a quantity that multiplies a vector to give another vector. It has magnitude but not direction.

For example:

$$s = \binom{2}{-2}$$

$$\therefore 2s = 2\binom{2}{-2} = \binom{4}{-4}$$

Multiplying the vectors by 2 doubles the length. The direction remains the same, so the vectors are parallel.

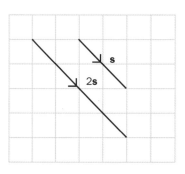

# Addition and subtraction of vectors

**KEY POINT**

Addition and subtraction of vectors produces a further vector called the **resultant**.

## Addition of vectors

- **Finding the resultant**

  When two vectors are added, the resultant vector is found by adding the horizontal units and then the vertical units:

  $$\begin{pmatrix} a \\ b \end{pmatrix} + \begin{pmatrix} c \\ d \end{pmatrix} = \begin{pmatrix} a + c \\ b + d \end{pmatrix}$$

  **Example**

  Find the resultant vector $\overrightarrow{CE}$ when $\overrightarrow{CD} = \begin{pmatrix} 3 \\ 4 \end{pmatrix}$ and $\overrightarrow{DE} = \begin{pmatrix} 2 \\ -1 \end{pmatrix}$

  $$\begin{aligned} \overrightarrow{CE} &= \overrightarrow{CD} + \overrightarrow{DE} \\ &= \begin{pmatrix} 3 \\ 4 \end{pmatrix} + \begin{pmatrix} 2 \\ -1 \end{pmatrix} \\ &= \begin{pmatrix} 3 + 2 \\ 4 + -1 \end{pmatrix} = \begin{pmatrix} 5 \\ 3 \end{pmatrix} \end{aligned}$$

- **Triangle law**
  Adding $\overrightarrow{PQ}$ and $\overrightarrow{QR}$ gives the resultant $\overrightarrow{PR}$.
  These vectors form a triangle.
  This can be written as **p** + **q** = **r**

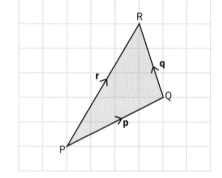

- **Parallelogram law**
  If the triangle above is developed into a parallelogram:
  $\overrightarrow{PQ} + \overrightarrow{QR} = \overrightarrow{PR}$ and $\overrightarrow{PS} + \overrightarrow{SR} = \overrightarrow{PR}$
  So **p** + **q** = **r** whichever way round the parallelogram is taken.
  i.e. **p** + **q** = **q** + **p** = **r**

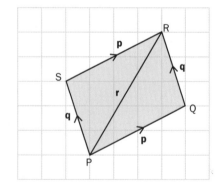

## Subtraction of vectors

In order to move from P to R it is necessary to
use the inverse vector.
The inverse of $\overrightarrow{QR}$ is $\overrightarrow{RQ}$ or $-\overrightarrow{QR}$.
$\overrightarrow{RQ} = \mathbf{q}$
$\mathbf{p} + -\mathbf{q} = \mathbf{r}$
$\mathbf{p} - \mathbf{q} = \mathbf{r}$

---

### Example

ABCD is a quadrilateral.
M, N and P are midpoints as shown.
$\overrightarrow{AB} = \mathbf{a}$, $\overrightarrow{CB} = \mathbf{c}$, $\overrightarrow{AD} = \mathbf{b}$, $\overrightarrow{DC} = \mathbf{d}$
Find **(a)** $\overrightarrow{AC}$ **(b)** $\overrightarrow{PA}$ **(c)** $\overrightarrow{MP}$ **(d)** $\overrightarrow{AM}$ **(e)** $\overrightarrow{NP}$

**(a)** $\overrightarrow{AC} = \overrightarrow{AD} + \overrightarrow{DC} = \mathbf{b} + \mathbf{d}$
or $\overrightarrow{AC} = \overrightarrow{AB} + \overrightarrow{BC} = \mathbf{a} - \mathbf{c}$

**(b)** $\overrightarrow{PA} = \overrightarrow{PB} + \overrightarrow{BA} = \frac{1}{2}\overrightarrow{CB} + (-\overrightarrow{AB}) = \frac{1}{2}\mathbf{c} - \mathbf{a}$

P is midpoint of $\overrightarrow{CB}$, so is half CB

Inverse vector

**(c)** $\overrightarrow{MP} = \frac{1}{2}\overrightarrow{DC} + \frac{1}{2}\overrightarrow{CB} = \frac{1}{2}\mathbf{d} + \frac{1}{2}\mathbf{c} = \frac{1}{2}(\mathbf{d} + \mathbf{c})$ — Common factor $\frac{1}{2}$

**(d)** $\overrightarrow{AM} = \overrightarrow{AD} + \overrightarrow{DM} = \mathbf{b} + \frac{1}{2}\mathbf{d}$

**(e)** $\overrightarrow{NP} = \overrightarrow{ND} + \overrightarrow{DC} + \overrightarrow{CP} = \frac{1}{2}\mathbf{b} + \mathbf{d} + \frac{1}{2}\mathbf{c}$
or $\overrightarrow{NP} = \overrightarrow{NA} + \overrightarrow{AB} + \overrightarrow{BP} = (-\frac{1}{2}\mathbf{d}) + \mathbf{a} + (-\frac{1}{2}\mathbf{c}) = \frac{-\mathbf{d}}{2} + \mathbf{a} - \frac{\mathbf{c}}{2} = \mathbf{a} - \frac{1}{2}(\mathbf{c} + \mathbf{d})$

---

# Real-life problems using vectors

| | |
|---|---|
| AQA UNITISED | ✓ |
| AQA LINEAR | ✓ |
| EDEXCEL A | ✓ |
| EDEXCEL B | ✓ |
| OCR A | ✓ |
| OCR B | ✓ |
| WJEC UNITISED | ✗ |
| WJEC LINEAR | ✗ |
| CCEA | ✗ |

Vectors are used to represent real-life problems involving forces such as river current, wind speed, etc. Swimming or sailing across a river are common vector problems.

It is always a good idea to adjust the given information into known vectors and use a vector triangle or parallelogram.

**Examples**

Jasmine swims across a river at 2.5m/s. The current is flowing at 2m/s.

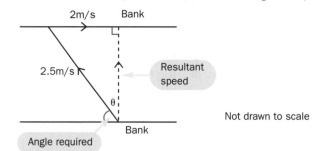

Not drawn to scale

**(a)** At what angle to the bank will she have to swim to reach the opposite bank at a right angle?

**(b)** What will her actual speed be?

**(a)** $\sin \theta = \dfrac{\text{opp}}{\text{hyp}}$

$\sin \theta = \dfrac{2}{2.5}$

$\therefore \theta = 53.1°$  ←  Use $\sin^{-1}$ key

Angle required = $90° - 53.1° = 36.9°$

**(b)** Resultant speed = $\sqrt{2.5^2 - 2^2}$  ←  Use Pythagoras' theorem

$= \sqrt{6.25 - 4}$

$= \sqrt{2.25} = 1.5\text{m/s}$

---

**PROGRESS CHECK**

1. Find:

   **(a)** $\mathbf{e} + \mathbf{f}$  **(b)** $\mathbf{e} - \mathbf{f}$  **(c)** $\mathbf{e} + \mathbf{f} + \mathbf{g}$  **(d)** $\mathbf{g} - \mathbf{e}$  when

   $\mathbf{e} = \begin{pmatrix} -3 \\ 2 \end{pmatrix}$     $\mathbf{f} = \begin{pmatrix} 2 \\ 1 \end{pmatrix}$     $\mathbf{g} = \begin{pmatrix} 3 \\ 4 \end{pmatrix}$

2. Four points are plotted and joined to form a quadrilateral.

   P (-8, 2); Q (-4, -3); R(8, 3); S(4, 8)

   **(a)** What are the column vectors for $\overrightarrow{PQ}$, $\overrightarrow{QR}$, $\overrightarrow{PS}$, $\overrightarrow{SR}$?

   **(b)** What shape is formed?

   **(c)** What can you say about the opposite sides of PQRS?

   **(d)** M is the midpoint of $\overrightarrow{RS}$. If $\overrightarrow{QR} = \mathbf{c}$ and $\overrightarrow{QP} = \mathbf{b}$, show that

   $\overrightarrow{QM} = \mathbf{c} + \dfrac{1}{2}\mathbf{b}$ whichever direction is taken to reach M.

   1. (a) $\begin{pmatrix} -1 \\ 3 \end{pmatrix}$ (b) $\begin{pmatrix} -5 \\ 1 \end{pmatrix}$ (c) $\begin{pmatrix} 2 \\ 7 \end{pmatrix}$ (d) $\begin{pmatrix} 6 \\ 2 \end{pmatrix}$  2. (a) $\overrightarrow{PQ} = \begin{pmatrix} 4 \\ -5 \end{pmatrix}$; $\overrightarrow{QR} = \begin{pmatrix} 12 \\ 6 \end{pmatrix}$; $\overrightarrow{PS} = \begin{pmatrix} 12 \\ 6 \end{pmatrix}$; $\overrightarrow{SR} = \begin{pmatrix} 4 \\ -5 \end{pmatrix}$
   (b) Parallelogram (c) Opposite sides are parallel and equal. (d) $\overrightarrow{QM} = \mathbf{c} + \dfrac{1}{2}\mathbf{b}$ or $\overrightarrow{QM} = \mathbf{b} + \mathbf{c} - \dfrac{1}{2}\mathbf{b}$

# 3.10 Constructions and loci

**LEARNING SUMMARY**

After studying this section, you should be able to understand:

- drawing lines and angles
- drawing triangles
- drawing quadrilaterals and inscribed polygons
- constructing loci

## Drawing lines and angles

AQA UNITISED ✓
AQA LINEAR ✓
EDEXCEL A ✓
EDEXCEL B ✓
OCR A ✓
OCR B ✓
WJEC UNITISED ✓
WJEC LINEAR ✓
CCEA ✓

### Lines

> Always use a sharp pencil for drawing accurate constructions. Never draw with pen in case you make a mistake.

**Example**

**(a)** Draw a line AB measuring 2.5cm.

**(b)** At A draw a line AC at an angle of 65°.

**(c)** Draw a line parallel to AC to cut AB at D.

**(a)** Use a ruler to draw a line and measure a length of 2.5cm. Mark this length AB.

**(b)** Use a protractor to measure an angle of 65° at A. Draw line AC.

**(c)** There are three ways to draw a line parallel to AC.

- Mark a point D on AB and measure an angle of 65° at D. This line will be parallel to AC because of the corresponding angles.
- Place a set square so that its hypotenuse is against AC. Place a straight edge along another side. Slide the set square along the straight edge and draw a line along the hypotenuse to cut AB at D. This line is parallel to AC.
- Use a pair of compasses with point on A and draw an arc to cut across AC at X and AB at D. With the same compass width and point on X, draw another arc to cross AC at Y and then with point on D draw an arc to cross upper arc at E. Draw a line through D and E. This line is parallel to AC.

> Always leave the arcs on your drawing. They show your method of construction.

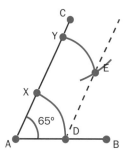

## Angles

Some angles can be constructed using only a ruler and a pair of compasses.

- **60°**

  If you are given length AB, use this as the compass width. Otherwise use any length.

  **1** Draw line and mark point A at one end.
  **2** Use a pair of compasses to draw arc cutting line at B.
  **3** With same compass width and point on B, draw a second arc to cross the first arc at C.
  **4** Join A to C.

  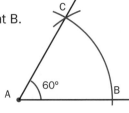

- **30°**

  **1** First draw an angle of 60° (as above).
  **2** With same compass width and point first on B and then C, draw arcs to cross at X.
  **3** Join A to X. This is the **bisector** of ∠ BAC.

  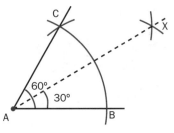

- **90°**

  **1** Draw line and mark point A at one end.
  **2** Draw a circle, centre O, to pass through A and cut line at B.
  **3** Draw line from B, through O, to circumference at C.
  **4** BC is the diameter of the circle, centre O. Join A to C. ∠ BAC = 90° (Angle in semicircle)

  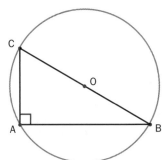

- **45°**

  First draw an angle of 90° as above and then bisect to form 45°

  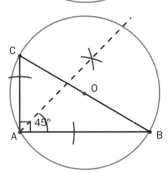

- **Drawing a perpendicular from point P to a line**

  **1** With compass point on P, draw arcs to cut line at Q and R.
  **2** With same compass width and point on Q and then R, draw arcs to cross below line.
  **3** Join crossed arcs to P to form perpendicular.

  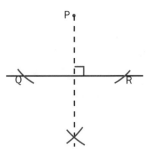

# Drawing triangles

Triangles can be drawn using a ruler, protractor and a pair of compasses.

- **When three sides are given**
  1. Draw line XY and measure to given length.
  2. With compass width equal to XZ and point on X, draw an arc.
  3. With compass width equal to YZ and point on Y, draw a second arc.

- **When two sides and an angle are given**
  1. Draw line XY and measure to given length.
  2. Use a protractor to measure and draw angle Y.
  3. Measure length YZ and mark Z.
  4. Join XZ.

- **When two angles and a side are given**
  1. Draw line XY and measure to given length.
  2. Use a protractor to measure and draw given angles at X and Y.
  3. Z is the point where these lines meet.

> This only works if the given side lies between the given angles.

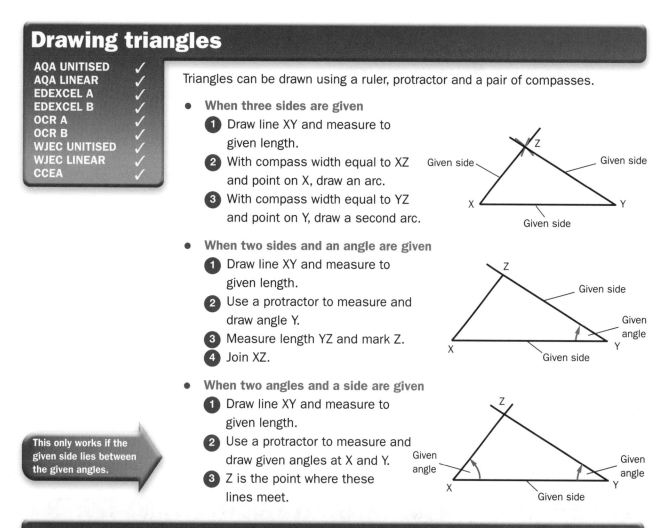

# Drawing quadrilaterals and inscribed polygons

## Drawing quadrilaterals

Drawing a quadrilateral is just an extension of drawing triangles.

> Draw a rough sketch first.

### Example

Construct a quadrilateral ABCD with AB = 6.5cm, AC = 5cm, BD = 5.8cm
∠ BDC = 100°, ∠ ABD = 72°,
Measure CD and ∠ BAC

- Draw line AB to given length.
- Measure and draw angle 72° at B.
- Measure length BD and mark D.
- Measure and draw angle 100° at D.
- With a pair of compasses, draw an arc of 5cm, centre A, crossing line from D. Mark this point C. Join AC.
- Measure CD = 4cm, ∠ BAC = 82°

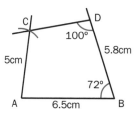

Not drawn to scale

## Inscribed polygons

An **inscribed polygon** is a polygon drawn in a circle with all vertices on the circumference.

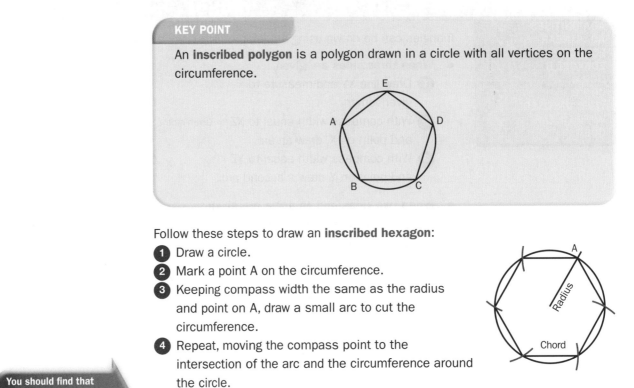

Follow these steps to draw an **inscribed hexagon**:

1. Draw a circle.
2. Mark a point A on the circumference.
3. Keeping compass width the same as the radius and point on A, draw a small arc to cut the circumference.
4. Repeat, moving the compass point to the intersection of the arc and the circumference around the circle.
5. Join each pair of points with a chord so that a hexagon is formed.

> You should find that each chord (side of the hexagon) equals the radius.

# Constructing loci

| | |
|---|---|
| AQA UNITISED | ✓ |
| AQA LINEAR | ✓ |
| EDEXCEL A | ✓ |
| EDEXCEL B | ✓ |
| OCR A | ✓ |
| OCR B | ✓ |
| WJEC UNITISED | ✓ |
| WJEC LINEAR | ✓ |
| CCEA | ✓ |

A **locus** is the path of a point that moves according to a given rule. Loci is the plural of locus.

- The locus of a point, moving at a fixed distance r from a fixed point C is a circle. Draw circle, radius $r$, centre, C.

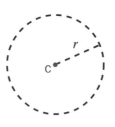

- The locus of a point, moving at a constant distance from a line through two fixed points A and B, is a pair of straight lines parallel to AB and at a distance $d$ from AB.
  1. Draw line AB.
  2. Measure distance $d$ above and below A and B.
  3. Join points to form two parallel lines.

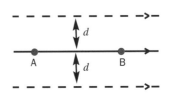

- The locus of a point, equidistant from two fixed points A and B, is the **perpendicular bisector** of AB.
  1️⃣ Draw line AB.
  2️⃣ Using compass width over half the length of AB and centres A and B, draw arcs to cross above and below AB.
  3️⃣ Join these cross points to form the perpendicular bisector cutting AB in half at right angles.

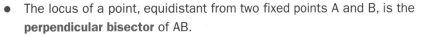

- The locus of a point equidistant from two fixed lines, is the bisector of the angle between the lines.
  1️⃣ Draw line XZ.
  2️⃣ Measure ∠ YXZ with a protractor.
  3️⃣ With compass point on X, draw arcs to cut XY and XZ.
  4️⃣ Keeping same compass width and compass point on the points of intersection of the arcs and lines, draw arcs to cross between the lines.
  5️⃣ Join crossed arcs to X to form angle bisector.

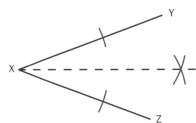

**PROGRESS CHECK**

1️⃣ Use only a ruler and a pair of compasses for this question.
 **(a)** Construct triangle PQR with PQ = 4.5cm, QR = 6cm, PR = 5.5cm
 **(b)** Bisect all three angles. Mark the point of intersection, X.
 **(c)** Draw a circle, centre X, with circumference touching all three sides.
 **(d)** Measure PX, RX and QX.

2️⃣ A market gardener is pestered by birds eating his fruit berries. A straight path of 8km (PR) runs through the middle of his fruit bushes. He decides to place scarecrows, at a variable point S, to frighten the birds away. S is always a distance of 5km from the midpoint (M) of PR and 4km from PR. By constructing the loci of S satisfying these conditions, find the positions that the market gardener can place his scarecrows.

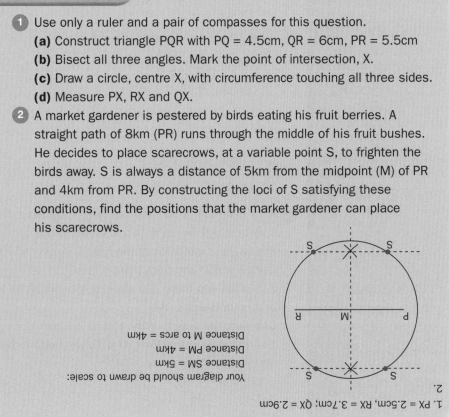

1. PX = 2.5cm, RX = 3.7cm; QX = 2.9cm

2.
Your diagram should be drawn to scale:
Distance SM = 5km
Distance PM = 4km
Distance M to arcs = 4km

# Sample GCSE questions

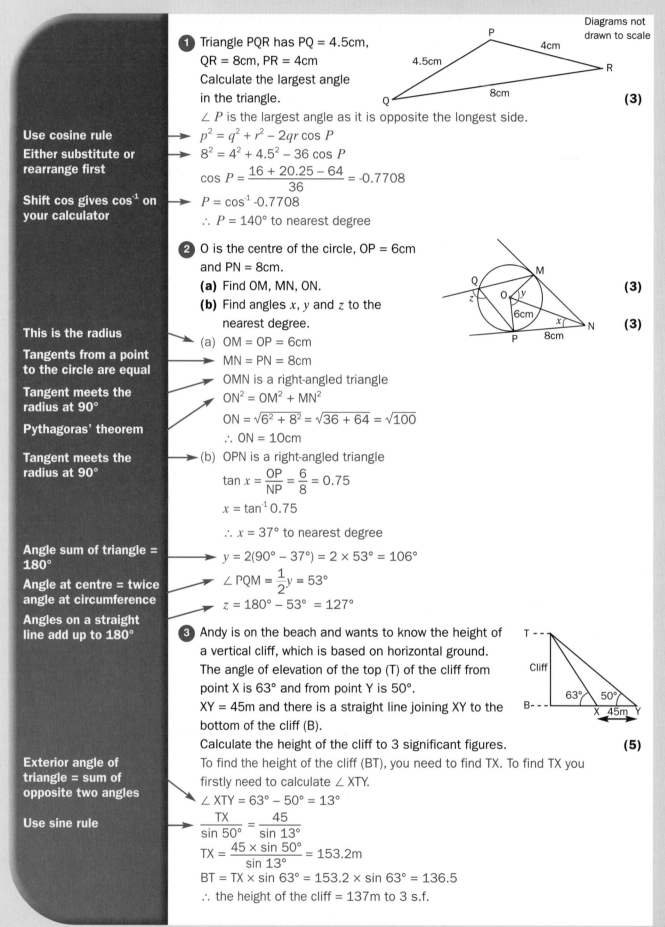

Diagrams not drawn to scale

**1** Triangle PQR has PQ = 4.5cm, QR = 8cm, PR = 4cm
Calculate the largest angle in the triangle. **(3)**

∠ P is the largest angle as it is opposite the longest side.

**Use cosine rule** → $p^2 = q^2 + r^2 - 2qr \cos P$

**Either substitute or rearrange first** → $8^2 = 4^2 + 4.5^2 - 36 \cos P$

$\cos P = \dfrac{16 + 20.25 - 64}{36} = -0.7708$

**Shift cos gives cos⁻¹ on your calculator** → $P = \cos^{-1} -0.7708$

∴ P = 140° to nearest degree

**2** O is the centre of the circle, OP = 6cm and PN = 8cm.
(a) Find OM, MN, ON. **(3)**
(b) Find angles $x$, $y$ and $z$ to the nearest degree. **(3)**

**This is the radius** → (a) OM = OP = 6cm

**Tangents from a point to the circle are equal** → MN = PN = 8cm

**Tangent meets the radius at 90°** → OMN is a right-angled triangle

**Pythagoras' theorem** → $ON^2 = OM^2 + MN^2$

$ON = \sqrt{6^2 + 8^2} = \sqrt{36 + 64} = \sqrt{100}$

∴ ON = 10cm

**Tangent meets the radius at 90°** → (b) OPN is a right-angled triangle

$\tan x = \dfrac{OP}{NP} = \dfrac{6}{8} = 0.75$

$x = \tan^{-1} 0.75$

∴ x = 37° to nearest degree

**Angle sum of triangle = 180°** → $y = 2(90° - 37°) = 2 \times 53° = 106°$

**Angle at centre = twice angle at circumference** → $\angle PQM = \dfrac{1}{2}y = 53°$

**Angles on a straight line add up to 180°** → $z = 180° - 53° = 127°$

**3** Andy is on the beach and wants to know the height of a vertical cliff, which is based on horizontal ground.
The angle of elevation of the top (T) of the cliff from point X is 63° and from point Y is 50°.
XY = 45m and there is a straight line joining XY to the bottom of the cliff (B).
Calculate the height of the cliff to 3 significant figures. **(5)**

To find the height of the cliff (BT), you need to find TX. To find TX you firstly need to calculate ∠ XTY.

**Exterior angle of triangle = sum of opposite two angles** → ∠ XTY = 63° − 50° = 13°

**Use sine rule** → $\dfrac{TX}{\sin 50°} = \dfrac{45}{\sin 13°}$

$TX = \dfrac{45 \times \sin 50°}{\sin 13°} = 153.2m$

BT = TX × sin 63° = 153.2 × sin 63° = 136.5

∴ the height of the cliff = 137m to 3 s.f.

# Exam practice questions

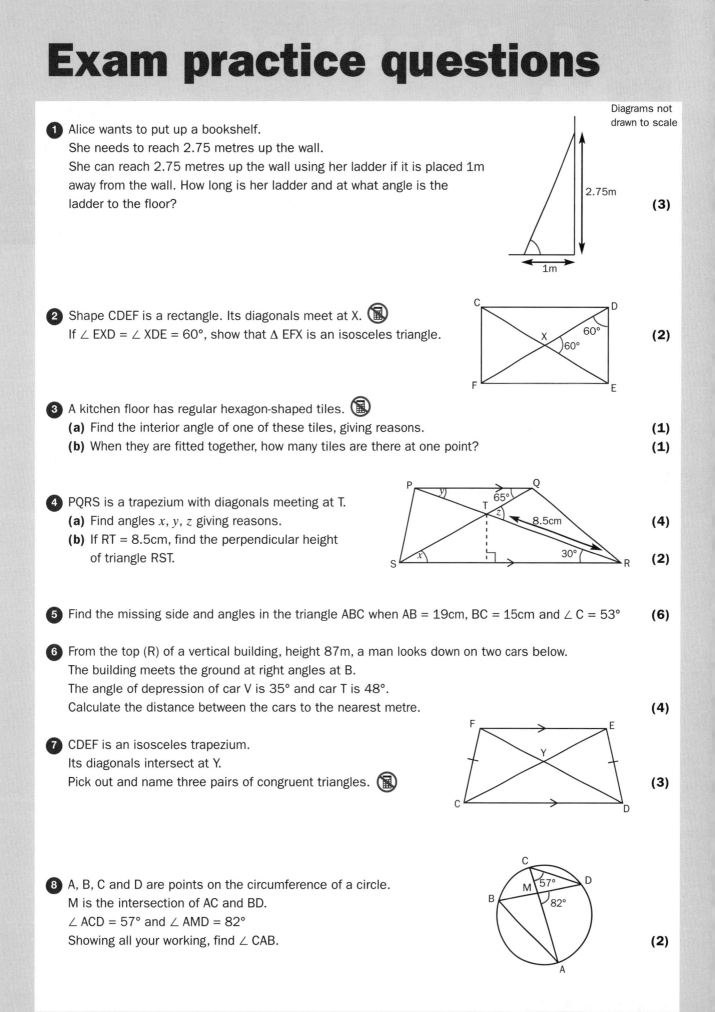

Diagrams not drawn to scale

**1** Alice wants to put up a bookshelf.
She needs to reach 2.75 metres up the wall.
She can reach 2.75 metres up the wall using her ladder if it is placed 1m
away from the wall. How long is her ladder and at what angle is the
ladder to the floor?   **(3)**

2.75m

1m

**2** Shape CDEF is a rectangle. Its diagonals meet at X.
If ∠ EXD = ∠ XDE = 60°, show that Δ EFX is an isosceles triangle.   **(2)**

**3** A kitchen floor has regular hexagon-shaped tiles.
(a) Find the interior angle of one of these tiles, giving reasons.   **(1)**
(b) When they are fitted together, how many tiles are there at one point?   **(1)**

**4** PQRS is a trapezium with diagonals meeting at T.
(a) Find angles $x$, $y$, $z$ giving reasons.   **(4)**
(b) If RT = 8.5cm, find the perpendicular height
of triangle RST.   **(2)**

**5** Find the missing side and angles in the triangle ABC when AB = 19cm, BC = 15cm and ∠ C = 53°   **(6)**

**6** From the top (R) of a vertical building, height 87m, a man looks down on two cars below.
The building meets the ground at right angles at B.
The angle of depression of car V is 35° and car T is 48°.
Calculate the distance between the cars to the nearest metre.   **(4)**

**7** CDEF is an isosceles trapezium.
Its diagonals intersect at Y.
Pick out and name three pairs of congruent triangles.   **(3)**

**8** A, B, C and D are points on the circumference of a circle.
M is the intersection of AC and BD.
∠ ACD = 57° and ∠ AMD = 82°
Showing all your working, find ∠ CAB.   **(2)**

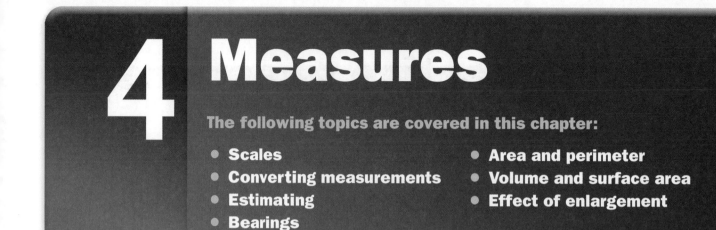

# 4 Measures

The following topics are covered in this chapter:

- Scales
- Converting measurements
- Estimating
- Bearings
- Area and perimeter
- Volume and surface area
- Effect of enlargement

# 4.1 Scales

<table>
<tr><td>LEARNING SUMMARY</td><td>After studying this section, you should be able to understand:<br><br>• maps and scale drawing<br>• instrument scales</td></tr>
</table>

## Maps and scale drawing

| | |
|---|---|
| AQA UNITISED | ✓ |
| AQA LINEAR | ✓ |
| EDEXCEL A | ✓ |
| EDEXCEL B | ✓ |
| OCR A | ✓ |
| OCR B | ✓ |
| WJEC UNITISED | ✓ |
| WJEC LINEAR | ✓ |
| CCEA | ✓ |

Maps and plans have to be drawn to scale to enable large distances to be shown on a page.

### Maps

> **KEY POINT**
>
> Maps are drawn to scale $1 : n$ where $n$ is the distance represented by 1cm.

**Examples**

Four towns C, D, H and J are in the same area. The distances between town G and these towns are shown on a map with scale 1 : 1 000 000.

**(a)** On the map, G → C = 6.3cm. What is the actual distance between towns G and C?

**(b)** The actual distance between G → H = 21.04km. What is the map distance between towns G and H?

**(a)** G → C = 6.3cm
Actual distance = 6.3 × 1 000 000      ← Multiply by scale factor
= 6 300 000cm
= 63 000m      ← Divide by 100 as 100cm = 1m
= 63km      ← Divide by 1000 as 1000m = 1km

**(b)** G → H = 21.04km
Map distance = $\dfrac{21.04 \times 100\,000}{1\,000\,000}$      ← Multiply by 100 000 to change km to cm

= 2.104cm      ← Divide by scale factor
= 2.1cm (1 d.p.)

> 1 d.p. is probably the most accurate measurement that can be drawn.

## Plans

A plan may be drawn to show an area of streets or the rooms in a building.

> **KEY POINT**
>
> A grid reference gives the horizontal and vertical readings.

**Examples**

What are the grid references of the doctor's, dentist's, chemist, station, school, village pond and park?

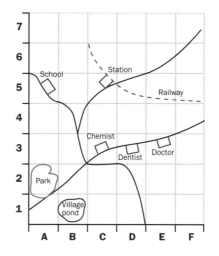

Read the grid by giving a letter and a number.

Doctor's E3, Dentist's D3, Chemist C3, Station C5, School A5, Village pond B1, Park A1/A2

## Scale drawing

Any length can be drawn to scale to enable a diagram to fit on a page. Divide each actual measurement by the scale to find the length to be drawn.

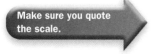

Make sure you quote the scale.

First put the given measurements on a rough sketch. If you are given a scale use it. If not, choose a scale convenient to use, which allows your drawing to fit on a page. A scale that produces too small a drawing will result in inaccuracies.

**Examples**

Use a scale of 1 : 20 to draw lines representing the following distances:
**(a)** 28cm **(b)** 1.9m

**Draw to 1 d.p.**

**(a)** 28cm ÷ 20 = 1.4cm ← Divide by scale

**(b)** 1.9m = 190cm ← Multiply by 100 to convert to cm
190cm ÷ 20 = 9.5cm

# Instrument scales

In our everyday life we need to read the scales on numerous instruments. Many instruments now have a digital read-out, but it is still necessary to know how to read off a dial or meter.

## Protractor

The 0° – 180° and 90° lines meet at the point of an angle. Place the protractor so that the base line of the angle is underneath it. The point of the angle should be under the meeting points of the lines on the protractor. There are two scales. Read the scale that has 0° on your angle base line.

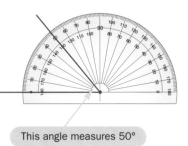

This angle measures 50°

## Other instruments

Read the scales on the following instruments:

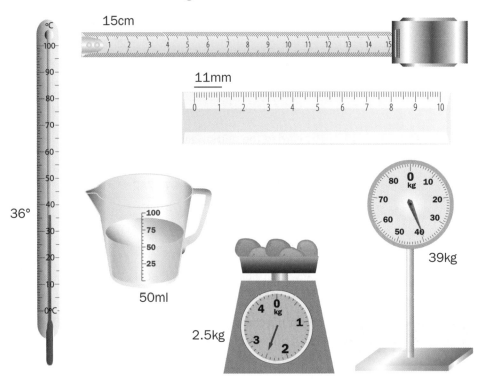

## Accuracy

It is important to realise that scales are often inaccurate. Reading the scale on a measuring tool has to be taken to be within certain limits. It is usual to give a length or an angle to 1 decimal place as it is impossible to read more decimal places.

Digital read-outs give more accurate measurements.

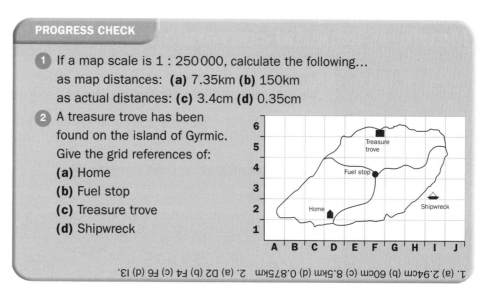

1 If a map scale is 1 : 250 000, calculate the following...
   as map distances: **(a)** 7.35km **(b)** 150km
   as actual distances: **(c)** 3.4cm **(d)** 0.35cm

2 A treasure trove has been found on the island of Gyrmic.
   Give the grid references of:
   **(a)** Home
   **(b)** Fuel stop
   **(c)** Treasure trove
   **(d)** Shipwreck

1. (a) 2.94cm (b) 60cm (c) 8.5km (d) 0.875km  2. (a) D2 (b) F4 (c) F6 (d) I3.

# 4.2 Converting measurements

| LEARNING SUMMARY | After studying this section, you should be able to understand: |
|---|---|
| | • metric → metric conversion |
| | • metric ↔ imperial conversion |
| | • units of time |
| | • compound measures |

## Metric → metric conversion

AQA UNITISED ✓
AQA LINEAR ✓
EDEXCEL A ✓
EDEXCEL B ✓
OCR A ✓
OCR B ✓
WJEC UNITISED ✓
WJEC LINEAR ✓
CCEA ✓

Measurements are given in the **metric system**. For example, mm, cm, g, l.

**KEY POINT**

Converting one metric unit to another metric unit involves multiplying or dividing by a power of 10.

You must learn conversion facts such as:

- 10mm = 1cm
- 100cm = 1m
- 1000m = 1km
- 1000mg = 1g
- 1000g = 1kg
- 1000kg = 1 tonne
- 1000ml = 1 litre
- 100cl = 1 litre
- 1000cm³ = 1 litre

Remember that prefix kilo- and milli- mean 1000 and centi- means 100.

Multiply if the answer number is going to be larger. Divide if the answer number is going to be smaller.

Remember that if 1m = 100cm, then 1m² = 100 × 100 = 10 000cm² and 1m³ = 100 × 100 × 100 = 1 000 000cm³

**Examples**

Change each unit to the one given in brackets.

**(a)** 0.628kg (g)
$$0.628\text{kg} = 0.628 \times 1000\text{g} \quad\longleftarrow\quad 1\text{kg} = 1000\text{g}$$
$$= 628\text{g}$$

**(b)** 35 000cm² (m²)
$$35\,000\text{cm}^2 = 35\,000 \div 10\,000\text{m}^2 \quad\longleftarrow\quad 1\text{m}^2 = 100\text{cm} \times 100\text{cm}$$
$$= 3.5\text{m}^2$$

# Metric ↔ imperial conversion

| AQA UNITISED | ✓ |
| AQA LINEAR | ✓ |
| EDEXCEL A | ✓ |
| EDEXCEL B | ✓ |
| OCR A | ✓ |
| OCR B | ✗ |
| WJEC UNITISED | ✓ |
| WJEC LINEAR | ✓ |
| CCEA | ✓ |

Measurements were previously given in the **imperial system**. Some of these units are still in use. For example, feet (ft), pint (pt), pounds (lbs), miles.

To change between metric and imperial systems you must learn some appropriate facts.

Approximate metric to imperial conversions:

> It is also useful to know that:
>
> 1lb = 16 ounces (oz)
> 1 yard (yd) = 3 feet (ft)
> 1 foot (ft) = 12 inches (ins)

- **Length**
  8km ≈ 5 miles    2.5cm ≈ 1 inch    30cm ≈ 1 foot    1m ≈ 1 yard (yd)

- **Area**
  1 hectare (ha) ≈ 2.5 acres ⟵ 1 acre = 4840 square yards
  1 hectare = 10 000m$^2$

- **Volume (capacity)**
  1 litre (l) ≈ 1.75 pints    4.5l ≈ 1 gallon

- **Mass (weight)**
  1kg ≈ 2.2lbs    1 tonne ≈ 1 ton ⟵ Note the different spellings
  1oz ≈ 28g

  Metric ↑    Imperial ↑

### Examples

Convert the following to the unit given in brackets.

**(a)** 3.5lbs (kg)
  3.5lbs ≈ 3.5 ÷ 2.2 ⟵ 1kg ≈ 2.2lbs
        ≈ 1.6kg

> Think about whether the answer is going to be larger or smaller.

**(b)** 53km (miles)
  $53\text{km} \approx 53 \times \frac{5}{8}$ miles ⟵ $1\text{km} \approx \frac{5}{8}$ miles
        ≈ 33 miles ⟵ Actual answer = 33.125, but this is an appropriate degree of accuracy

# Units of time

| AQA UNITISED | ✓ |
| AQA LINEAR | ✓ |
| EDEXCEL A | ✓ |
| EDEXCEL B | ✓ |
| OCR A | ✗ |
| OCR B | ✗ |
| WJEC UNITISED | ✓ |
| WJEC LINEAR | ✓ |
| CCEA | ✓ |

Time is measured in two different systems based on the 12 hour and 24 hour clocks.

> **KEY POINT**
>
> The **12 hour clock** registers time in two periods of 12 hours a.m. and p.m.
> The **24 hour clock** is a continuous time period of 24 hours.

## 12 hour clock

> **KEY POINT**
>
> Time period from midnight to noon is a.m.
> Time period from noon to midnight is p.m.

Time given in the 12 hour clock system must be followed by a.m. or p.m. to show whether it is before noon or after noon. For example, 6 a.m. is in the morning and 6 p.m. is in the evening.

## 24 hour clock

**12 hours are added to the morning time to give the afternoon time.**

### KEY POINT

The 24 hour clock covers the whole 24 hour period. Time is always given with four digits, for example 0600 is 6 o'clock in the morning and 1800 is 6 o'clock in the evening.

**Digital clocks give midnight as 0000 or 2400.**

You may need to convert times to the other time system, for example:

- 0700 → 7 a.m.
- 7 p.m. → 1900 ← Add 12 + 7 = 19
- 1625 → 4.25 p.m. ← Subtract 16 − 12 = 4

## Facts about time

It is important that you learn the following facts:

- 60 seconds (secs) = 1 minute (min)
- 60mins = 1 hour (hr)
- 24hrs = 1 day
- 7 days = 1 week

- 52 weeks = 1 year
- 365 days = 1 year (except in a leap year which has 366 days. A leap year occurs every 4th year.)

**Try learning the rhyme '30 days hath September, April, June and November...'**

Months have different number of days, usually 30 or 31. Check the calendar for these. February has 28 days in a regular year and 29 days in a leap year.

**Fewer mistakes occur when you use this method.**

**Do not subtract in the usual way. Remember that time is based on the number 60 not 10.**

### Example

How long is a TV programme starting at 7.05 p.m. and ending at 8.10 p.m?

7.05 p.m. → 8 p.m = 55mins          60mins = 1hr

7.05 p.m. → 8.10 p.m = 55mins + 10mins = 65mins or 1hr 5mins

# Compound measures

| AQA UNITISED | ✓ |
| AQA LINEAR | ✓ |
| EDEXCEL A | ✓ |
| EDEXCEL B | ✓ |
| OCR A | ✓ |
| OCR B | ✓ |
| WJEC UNITISED | ✓ |
| WJEC LINEAR | ✓ |
| CCEA | ✓ |

### KEY POINT

**Compound measures** use more than one unit. Examples are speed, density and rate of flow.

## Speed

**This formula triangle will help with calculations involving speed.**

Speed = Distance ÷ Time
Time = Distance ÷ Speed
Distance = Speed × Time

### KEY POINT

**Speed** = Distance ÷ Time

Speed can be measured in km/hr (kilometres per hour), mph (miles per hour) or m/s (metres per second). Speed gives the distance travelled in one unit of time.

### Example

What is the speed of a car taking $3\frac{1}{2}$ hours to drive 200 miles?

$$\text{Speed} = \frac{\text{Distance}}{\text{Time}} = \frac{200}{3.5}$$ ← Time in hours

$$= 57\text{mph (to nearest whole number)}$$ ← Miles per hour

## Average speed

**KEY POINT**

**Average speed** over a journey is found by the formula:

$$\text{Average speed} = \frac{\text{Total distance travelled}}{\text{Total time taken}}$$

### Example

A train travels 65 miles stopping 5 minutes at each of the 7 stations. It takes 1hr 20mins for the total journey. What is its average speed when travelling?

$$\text{Time taken for journey} = 1\text{hr } 20\text{mins} - \text{stopping time}$$
$$= 1\text{hr } 20\text{mins} - 35\text{mins} = 45\text{mins}$$

$$\text{Average speed} = \frac{\text{Total distance travelled}}{\text{Total time taken}}$$

$$= \frac{65 \text{ miles}}{45\text{mins}}$$

$$= \frac{65}{0.75}$$ ← Convert to hours

$$= 87\text{mph}$$

## Density

This formula triangle will help with calculations involving density.

Density = Mass ÷ Volume
Volume = Mass ÷ Density
Mass = Density × Volume

**KEY POINT**

$$\text{Density} = \frac{\text{Mass}}{\text{Volume}}$$

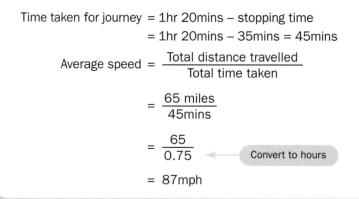

Density can be measured in $kg/m^3$ (kilograms per cubic metre) or $g/cm^3$ (grams per cubic centimetre).

### Example

Find the density of a block of stone weighing 850g with a volume of $92cm^3$.

$$\text{Density} = \frac{\text{Mass}}{\text{Volume}}$$

$$= \frac{850}{92}$$

$$= 9.24\text{g/cm}^3 \text{ (3 s.f.)}$$

# Other compound measures

> **KEY POINT**
>
> **Rate of flow** measures the amount of fluid that flows in a given time.
> Miles per gallon (mpg) gives the **fuel consumption** over a given distance.

### Example

What is the fuel consumption in mpg of a car travelling 200 miles and using 22 litres of fuel?

Fuel = 22l = 22 ÷ 4.5 gallons ← $4.5l \approx 1$ gallon

= 4.888... gallons

Fuel consumption = 200 ÷ 4.888 = 40.9mpg

**PROGRESS CHECK**

1. Convert these measurements to the unit given in brackets.
   (a) $25\,500cm^2$ $(m^2)$  (b) $20\,000m^2$ (ha)
   (c) $550mm^3$ $(cm^3)$  (d) 5.53l (ml)  (e) 33.4m (km)

2. Below is a recipe for biscuits, with quantities given in imperial measures. Rewrite the recipe in metric units. Do all the measures need to be changed?

   | | | |
   |---|---|---|
   | 3ozs margarine | 3ozs caster sugar | 2 tsp ginger |
   | 2 tbsp syrup | 8ozs self-raising flour | a pinch of salt |
   | $\frac{1}{2}$ tsp bicarbonate of soda | | |

3. This schedule gives TV programme times for an evening.

   | Time | Programme | Time | Programme |
   |------|-----------|------|-----------|
   | 5.55 | Local and national news | 8.05 | Live concert from WZ Arena |
   | 6.30 | Country Walks | 10.00 | Local and national news |
   | 6.55 | Beech Street | 10.35 | Film: Where Are We Now? |
   | 7.30 | Cook! Cook! Cook! | 12.25 | Football from the USA |

   (a) In which clock system are the times given?
   (b) Is there anything missing from the times?
   (c) How long is the film?
   (d) How long are you watching TV for, if you watch Beech Street and Cook! Cook! Cook!?
   (e) Convert all the times to the other clock system.

4. Calculate in given units:
   (a) Speed to travel 64 miles in 145mins   (mph to 3 s.f.)
   (b) Time taken to travel 290km at an average speed of 45mph (hrs to nearest min)
   (c) Density of a piece of rock of mass 3kg and volume $84cm^3$ ($g/cm^3$ to 3 s.f.)

4.(a) 26.5mph (b) 4hrs 2mins (c) $35.7g/cm^3$
(e) 1755, 1830, 1855, 1930, 2005, 2200, 2235, 0025
(c) 1hr 50mins (d) 1hr 10mins
3. (a) 12 hour clock (b) Times should have p.m. except for 12.25a.m.
2. 84g margarine, 84g caster sugar, 2 tbsp syrup, 224g self-raising flour,
½ tsp bicarbonate of soda, 2 tsp ginger, a pinch of salt; no
1. (a) $2.55m^2$ (b) 2ha (c) $0.55cm^3$ (d) 5530ml (e) 0.0334km

# 4.3 Estimating

**After studying this section, you should be able to understand:**

- estimating lines and angles
- estimating area and volume
- estimating weight

## Estimating lines and angles

| | |
|---|---|
| AQA UNITISED | ✓ |
| AQA LINEAR | ✓ |
| EDEXCEL A | ✓ |
| EDEXCEL B | ✓ |
| OCR A | ✓ |
| OCR B | ✓ |
| WJEC UNITISED | X |
| WJEC LINEAR | X |
| CCEA | ✓ |

Sometimes it is necessary to make a sensible estimate of a range of measures in real-life situations.

### Length

**Example**

Estimate the height of the bus.

The average estimated height of a man is just under 2 metres.

Judge how many times the picture of the man can fit into the height of the bus and multiply by 2.

It seems to be approximately 2.5 times, so the height of the bus is approximately 5 metres.

### Angles

**Examples**

Estimate the size of the following angles.

(a)   (b)   (c)

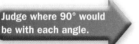

Judge where 90° would be with each angle.

(a) This is a reflex angle. The remaining part of the revolution looks as if it is just less than a right angle ∴ it is approximately 275°.

(b) This obtuse angle looks as if it is about 20° more than a right angle ∴ it is approximately 110°.

(c) This acute angle is less than half a right angle ∴ it is approximately 40°.

# Estimating area and volume

| | |
|---|---|
| AQA UNITISED | ✓ |
| AQA LINEAR | ✓ |
| EDEXCEL A | ✓ |
| EDEXCEL B | ✓ |
| OCR A | ✓ |
| OCR B | ✓ |
| WJEC UNITISED | ✗ |
| WJEC LINEAR | ✗ |
| CCEA | ✓ |

Estimating an area or volume involves estimating the dimensions first or counting squares or cubes in the shape on a 1cm square grid.

## Area

See page 134 for how to calculate area

Judge a length by measuring with the top joint of your thumb. This is approximately 2.5cm.

### Examples

**1.** Estimate the shaded area of this frame.

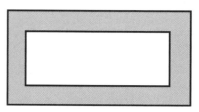

Taking a 'thumb' length, the outside rectangle measures roughly 5cm by 2.5cm. The border seems to be approximately 0.5cm wide.

Shaded area $\approx (5 \times 2.5) - (4 \times 1.5)$ — Subtract $(2 \times 0.5)$cm from each dimension

$= 12.5 - 6 = 6.5 \text{cm}^2$

**2.** Estimate the area of this shape drawn on a 1cm square grid.

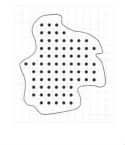

Count all fully included 1cm squares (●) = 74
Count all squares > one half = 23
Ignore all squares < one half
Total number of squares = 97
Estimated area of shape = 97cm²

## Volume

See page 138 for how to calculate volume

### Examples

**1.** The side of the square cross-section is 6.5m. Estimate the length and thus the volume of this cuboid.

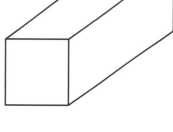

The length looks approximately double the given side so is approximately 13m.

Volume $\approx 6.5 \times 6.5 \times 13 = 549.25 \text{m}^3$
It is sensible to give the answer as 550m³.

**2.** Estimate the volume of this 3D shape made up of 1cm³ cubes.

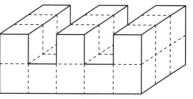

You need to visualise the cubes you cannot see.

You can see 8 cubes at the front of the shape. You can see 3 rows of cubes at the side of the shape.

Total number of cubes = 8 × 3 = 24
Estimated volume of shape = 24cm³

## Estimating weight

AQA UNITISED ✓
AQA LINEAR ✓
EDEXCEL A ✓
EDEXCEL B ✓
OCR A ✓
OCR B ✓
WJEC UNITISED ✗
WJEC LINEAR ✗
CCEA ✓

Weight often needs to be estimated in everyday situations. For example, the weight of loose fruit and vegetables is gauged if there are no scales at the supermarket.

### Example

How much does a loaf of bread weigh?

80g, 800g, 8kg or 80kg

The answer has to be sensible. You should realise that 8kg and 80kg are far too heavy. If you compare the loaf to a bag of sugar, which weighs 1kg or 1000g, then the loaf probably weighs 800g.

### PROGRESS CHECK

**Without using a calculator**

1 Estimate the following measures and give the appropriate units:
   (a) The length of a sofa that can seat three people
   (b) The height of a door
   (c) The area of a tennis court
   (d) The mass of a small car.
2 (a) A petrol tank has a capacity of 10 gallons. Approximately how many litres does it hold?
   (b) A water butt holds 200l of rain water. Estimate how many gallons it will hold.
3 Estimate the sizes of these angles.

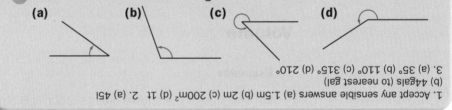

3. (a) 35° (b) 110° (c) 315° (d) 210°
(b) 44gals (to nearest gal)
1. Accept any sensible answers (a) 1.5m (b) 2m (c) 200m² (d) 1t   2. (a) 45l

# 4.4 Bearings

**LEARNING SUMMARY**

After studying this section, you should be able to understand:
● compass points
● three-figure bearings
● bearings diagrams

## Compass points

AQA UNITISED ✓
AQA LINEAR ✓
EDEXCEL A ✓
EDEXCEL B ✓
OCR A ✓
OCR B ✓
WJEC UNITISED ✓
WJEC LINEAR ✓
CCEA ✓

### KEY POINT

The four main **compass points** are North, South, East and West. The initial letters are usually used.

There is a right angle between:

- North and East
- East and South
- South and West
- West and North

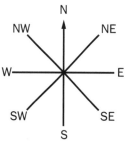

NE, SE, SW and NW points bisect each right angle. These angles are always measured clockwise from North. East is at 90°, South is at 180°, West is at 270°.

### Examples

Measuring clockwise from North...

**(a)** what is the angle between N and SE?

Angle = 3 × 45° = 135°

**(b)** what is the angle between NW and S?

Angle = 45° + 180° = 225°

# Three-figure bearings

AQA UNITISED ✓
AQA LINEAR ✓
EDEXCEL A ✓
EDEXCEL B ✓
OCR A ✓
OCR B ✓
WJEC UNITISED ✓
WJEC LINEAR ✓
CCEA ✓

**KEY POINT**

**Bearings** are used to describe directions. Bearings are measured using angles in a clockwise direction from the North line.

The direction of point B from A is given by the angle measured clockwise from North at point A and written as a **three-figure** angle.

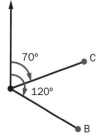

The bearing of B from A = 120°

The bearing of C from A = 070°  ← Use zero to make three figures

**KEY POINT**

If the bearing is given for a journey from X to Y, the **reverse bearing** gives the angle for the return journey (Y to X).

Reverse bearing is also known as the reciprocal or back bearing.

**Example**

Find the bearing from X to Y in each drawing and then find the bearing for the return journeys from Y to X.

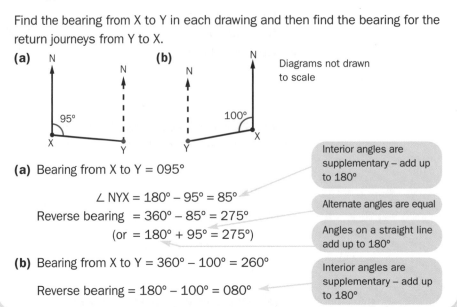

**(a)** Bearing from X to Y = 095°

$$\angle NYX = 180° - 95° = 85°$$
$$\text{Reverse bearing} = 360° - 85° = 275°$$
$$(\text{or} = 180° + 95° = 275°)$$

**(b)** Bearing from X to Y = 360° - 100° = 260°

$$\text{Reverse bearing} = 180° - 100° = 080°$$

> Interior angles are supplementary – add up to 180°

> Alternate angles are equal

> Angles on a straight line add up to 180°

> Interior angles are supplementary – add up to 180°

> It is necessary to draw North lines at Y. These are obviously parallel to the North lines at X. The bearing at Y is the clockwise angle from the North line to line YX.

# Bearings diagrams

Never assume diagrams are accurate unless told so.

## Given diagrams

You may be given a diagram of a journey and asked to calculate bearings or distances. Trigonometry or Pythagoras' theorem are often used.

**Example**

E and H are two small coastal towns. B is a village nearby. A lighthouse (L) is situated on the direct route from H to B, so that a perpendicular line joins it to E.

**(a)** Calculate the bearing of H from E, B from H and B from E.
**(b)** Calculate the distance from H to B.

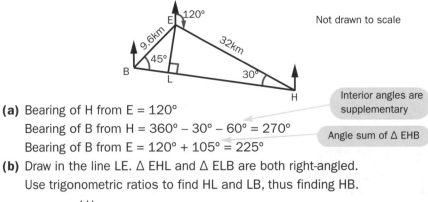

Not drawn to scale

**(a)** Bearing of H from E = 120°
Bearing of B from H = 360° - 30° - 60° = 270°
Bearing of B from E = 120° + 105° = 225°

> Interior angles are supplementary

> Angle sum of △ EHB

**(b)** Draw in the line LE. △ EHL and △ ELB are both right-angled.
Use trigonometric ratios to find HL and LB, thus finding HB.

In △ EHL, $\dfrac{LH}{32} = \cos 30°$

$\quad LH = 32 \times \cos 30° = 27.7\text{km (3 s.f.)}$

In △ ELB, $\dfrac{LB}{9.6} = \cos 45°$

$\quad LB = 9.6 \times \cos 45° = 6.79\text{km (3 s.f.)}$

∴ Distance HB = 27.7 + 6.79 = 34.49km = 34.5km (3 s.f.)

# Drawing scale diagrams

Drawing scale diagrams involves drawing lengths and angles to scale, as well as measuring other lengths and converting them back to actual distances.

> It is useful to draw a rough sketch first.

### Example

A ship sails 25km from port A on a bearing of 070° to port C. It then sails 30km due east to port B, before returning directly to port A. Use a scale drawing to find the distance and bearing of the return journey.

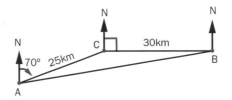

Choose a scale to fit the diagram on the page, e.g. scale 1cm : 10km

Place port A and draw in a North line. Measure 70° from North line clockwise. Measure 2.5cm (25km) and mark port C.

At port C, draw another North line. Measure 90° from North line clockwise. Measure 3cm (30km) and mark port B.

> N to E is a right angle

Join B to A and measure the distance with a ruler.

Actual distance = 5.4cm × 10km = 54km

Use a protractor to measure ∠ CBA = 9°

Bearing of A from B = 360° − 90° − 9° = 261°
(or = 180° + 81° = 261°)

### PROGRESS CHECK

**1** What is the bearing of P from R in each of these diagrams?

**(a)** **(b)** **(c)**

**2** A bird flies from its nest on a bearing of 050° to a tower 50km away. It then flies on a bearing of 160° to a roof 20km away. Use a scale drawing to find the bearing and journey distance for the bird's return to its nest.

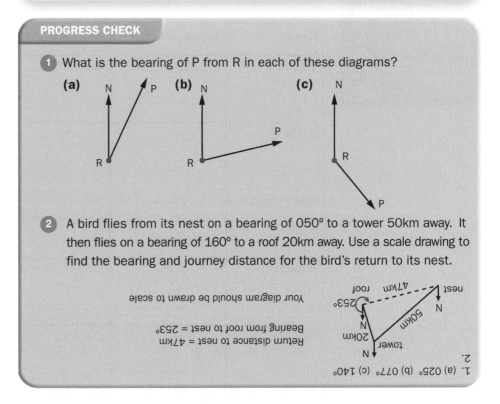

Your diagram should be drawn to scale
Bearing from roof to nest = 253°
Return distance to nest = 47km

1. (a) 025° (b) 077° (c) 140°

# 4.5 Area and perimeter

| | After studying this section, you should be able to understand: |
|---|---|
| **LEARNING SUMMARY** | • perimeter<br>• area and perimeter of 2D shapes<br>• area and perimeter of circles<br>• area and perimeter of compound shapes |

## Perimeter

AQA UNITISED ✓
AQA LINEAR ✓
EDEXCEL A ✓
EDEXCEL B ✓
OCR A ✓
OCR B ✓
WJEC UNITISED ✓
WJEC LINEAR ✓
CCEA ✓

**KEY POINT**

The **perimeter** of a shape is the distance all around the edge.

For example:

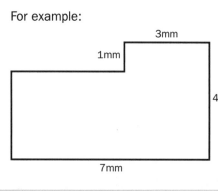

> Remember to count all the sides.

Perimeter = (7 + 4 + 3 + 1 + 4 + 3)mm
= 22mm

## Area and perimeter of 2D shapes

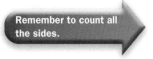

AQA UNITISED ✓
AQA LINEAR ✓
EDEXCEL A ✓
EDEXCEL B ✓
OCR A ✓
OCR B ✓
WJEC UNITISED ✓
WJEC LINEAR ✓
CCEA ✓

**Area** is always measured in squared units.

| Name | Shape | Perimeter | Area |
|---|---|---|---|
| Square | | $4a$ | $a^2$ |
| Rectangle | | $2(b + h)$ | $bh$ |
| Parallelogram | | $2(a + b)$<br>$a$ = slant height | $bh$<br>$h$ = perpendicular height |
| Trapezium | | $a + b + c + d$ | $\frac{1}{2}(a + b)h$ |
| Triangle | | $a + b + c$ | $\frac{1}{2}bh$<br><br>(half a parallelogram) |

### Examples

Calculate the areas of these shapes.

**(a)** 3cm  7.5cm

**(b)** 5cm  6.4cm

**(c)** 5cm  4.2cm  7cm

Diagrams not drawn to scale

**(a)** Area = $\frac{1}{2} \times 7.5 \times 3 = 11.25\text{cm}^2$

**(b)** Area = $6.4 \times 5 = 32\text{cm}^2$

**(c)** Area = $\frac{1}{2}(5 + 7) \times 4.2 = 25.2\text{cm}^2$

# Area of and perimeter of circles

AQA UNITISED ✓
AQA LINEAR ✓
EDEXCEL A ✓
EDEXCEL B ✓
OCR A ✓
OCR B ✓
WJEC UNITISED ✓
WJEC LINEAR ✓
CCEA ✓

**KEY POINT**

The perimeter of a circle is called the circumference ($C$). The formula for the perimeter of a circle is:

$C = 2\pi r$ or $C = \pi d$ ← Diameter ($d$) = 2 × radius ($r$)

The formula for the area of a circle ($A$) is:

$A = \pi r^2$

(Use $\pi = 3.142$ or $\pi$ key on the calculator. You may sometimes need to use $\pi = \frac{22}{7}$ or $\pi \approx 3$)

> Check if the radius or diameter is given in the question. Radius is needed to calculate area.

### Example

Find the circumference and area of a circle, radius = 3.4cm.

$C = 2\pi r$
$= 2\pi(3.4)$
$= 6.8\pi$
$= 21.4\text{cm}$ (3 s.f.)
$A = \pi r^2$
$= \pi(3.4)^2$
$= 36.3\text{cm}^2$ (3 s.f.)

## Length of an arc and area of a sector

An arc is part of the circumference of a circle. It subtends an angle at the centre of a circle. A sector is a section of a circle between two radii and an arc.

**KEY POINT**

If the angle is $\theta$ and the radius is $r$

Length of arc = $\frac{\theta}{360°} \times 2\pi r$

Area of sector = $\frac{\theta}{360°} \times \pi r^2$

> $= \frac{\theta}{360°}$ gives the fraction the sector is of the circle.

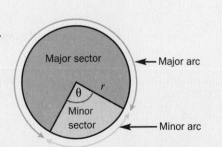

Major sector, Major arc, Minor sector, Minor arc, $\theta$, $r$

**Example**

**(a)** Find the length of the minor arc of a circle, diameter = 11cm, subtending an angle of 55° and **(b)** the area of its minor sector.

**(a)** Length of arc $= \dfrac{55}{360} \times \pi \times 11$ ⟵ $d = 11\text{cm}$

$= 5.28\text{cm}$

**(b)** Area of sector $= \dfrac{55}{360} \times \pi \times (5.5)^2$ ⟵ $r = 11 \div 2\text{cm}$

$= 14.5\text{cm}^2$

# Area of a segment

A segment is the area between a chord and the circumference.

> **KEY POINT**
>
> Area of minor segment = area of minor sector − Δ AOB
> Area of major segment = area of major sector + Δ AOB
>
>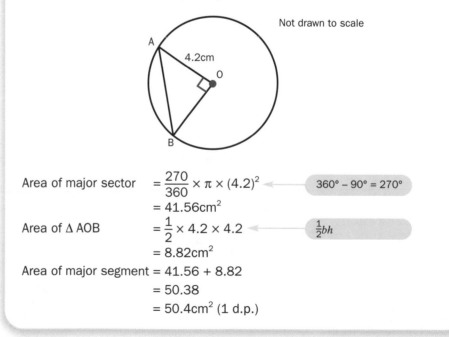

**Example**

Calculate the area of the major segment of this circle:

Not drawn to scale

Area of major sector $= \dfrac{270}{360} \times \pi \times (4.2)^2$ ⟵ $360° - 90° = 270°$

$= 41.56\text{cm}^2$

Area of Δ AOB $= \dfrac{1}{2} \times 4.2 \times 4.2$ ⟵ $\frac{1}{2}bh$

$= 8.82\text{cm}^2$

Area of major segment $= 41.56 + 8.82$

$= 50.38$

$= 50.4\text{cm}^2$ (1 d.p.)

# Area and perimeter of compound shapes

You may be asked to find the area and perimeter of a figure made up of different shapes.

### Example

A window is made up of glass and thin metal strips.
What area of glass and length of metal strip is required?

Not drawn to scale

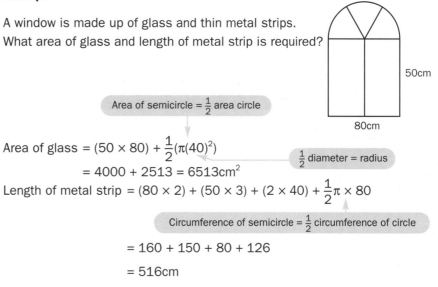

50cm

80cm

Area of semicircle = $\frac{1}{2}$ area circle

Area of glass = $(50 \times 80) + \frac{1}{2}(\pi(40)^2)$

= 4000 + 2513 = 6513cm$^2$

$\frac{1}{2}$ diameter = radius

Length of metal strip = $(80 \times 2) + (50 \times 3) + (2 \times 40) + \frac{1}{2}\pi \times 80$

Circumference of semicircle = $\frac{1}{2}$ circumference of circle

= 160 + 150 + 80 + 126

= 516cm

---

### PROGRESS CHECK

**1** These shapes are plans for flower beds in a park. Find the perimeters of **(a)** and **(d)**. Find the areas of **(b)** and **(c)**.

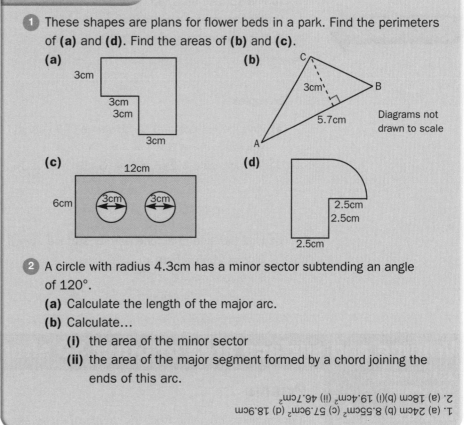

**(a)**
3cm
3cm
3cm
3cm

**(b)**
C
3cm
B
5.7cm
A

Diagrams not drawn to scale

**(c)**
12cm
6cm
3cm
3cm

**(d)**
2.5cm
2.5cm
2.5cm

**2** A circle with radius 4.3cm has a minor sector subtending an angle of 120°.

**(a)** Calculate the length of the major arc.

**(b)** Calculate...

   **(i)** the area of the minor sector

   **(ii)** the area of the major segment formed by a chord joining the ends of this arc.

2. (a) 18cm (b)(i) 19.4cm² (ii) 46.7cm²
1. (a) 24cm (b) 8.55cm² (c) 57.9cm² (d) 18.9cm

# 4.6 Volume and surface area

**LEARNING SUMMARY**

After studying this section, you should be able to understand:

- volume and surface area of cubes and cuboids
- volume and surface area of other 3D shapes
- volume and surface area of compound solid shapes

## Volume and surface area of cubes and cuboids

AQA UNITISED ✓
AQA LINEAR ✓
EDEXCEL A ✓
EDEXCEL B ✓
OCR A ✓
OCR B ✓
WJEC UNITISED ✓
WJEC LINEAR ✓
CCEA ✓

See Geometry 3.7 for different types of 3D shapes.

Imagine the 3D shape opened out into its net.

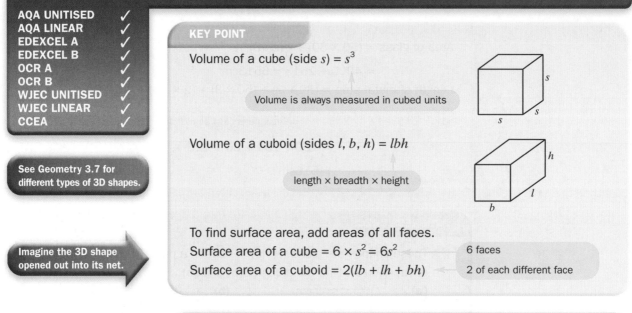

**KEY POINT**

Volume of a cube (side $s$) = $s^3$

Volume is always measured in cubed units

Volume of a cuboid (sides $l, b, h$) = $lbh$

length × breadth × height

To find surface area, add areas of all faces.
Surface area of a cube = $6 \times s^2 = 6s^2$    6 faces
Surface area of a cuboid = $2(lb + lh + bh)$    2 of each different face

**Examples**

**(a)** Find the volume and surface area of a cuboid $l$ = 6cm, $b$ = 4cm, $h$ = 3cm
Volume = $lbh$ = 6 × 4 × 3 = 72cm$^3$
Surface area = 2((6 × 4) + (6 × 3) + (4 × 3))
                = 2(24 + 18 + 12)
                = 2 × 54
                = 108cm$^2$

**(b)** Find the side of a cube and its surface area if its volume is 125cm$^3$.
Side = $\sqrt[3]{125}$ = 5cm
Surface area = 6 × 5$^2$
                = 150cm$^2$

## Volume and surface area of other 3D shapes

AQA UNITISED ✓
AQA LINEAR ✓
EDEXCEL A ✓
EDEXCEL B ✓
OCR A ✓
OCR B ✓
WJEC UNITISED ✓
WJEC LINEAR ✓
CCEA ✓

### Prisms

A prism takes its name from the shape of its cross-section.

**KEY POINT**

Volume of a prism = area of its cross-section × length (or height)

Surface area of a prism = sum of area of all faces

- **Triangular prism**

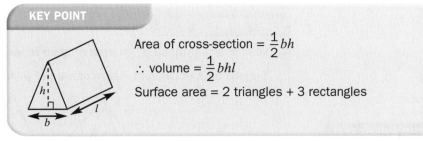

**KEY POINT**

Area of cross-section = $\frac{1}{2}bh$

∴ volume = $\frac{1}{2}bhl$

Surface area = 2 triangles + 3 rectangles

---

**Example**

Find the volume and surface area of a triangular prism where $h$ = 5cm, $b$ = 4cm and $l$ = 6cm.

Volume = $\frac{1}{2}(4 \times 5) \times 6 = 60$cm$^3$

To find the surface area, we need to find the side of the rectangle(s):

Using Pythagoras: $s^2 = 5^2 + 2^2$ ← $\frac{1}{2}$ base of triangle

$= 25 + 4 = 29$

$= \sqrt{29} = 5.39$cm

2 triangles + 2 rectangles + base

∴ surface area = $2(\frac{1}{2} \times 4 \times 5) + 2(5.39 \times 6) + (4 \times 6)$

$= 20 + 64.68 + 24 = 108.68$cm$^2$ = 109cm$^2$

Answers have been corrected to 3 s.f. or nearest whole number where appropriate.

---

- **Cylinder**

A cylinder is a prism with a circle as its cross-section.

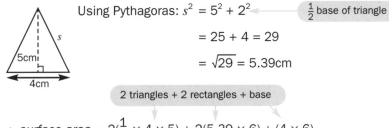

**KEY POINT**

Circle at end

Circumference of end

Volume = $\pi r^2 h$

To find the surface area, imagine it opened out to a net:

Surface area = $2\pi rh + 2\pi r^2$

---

**Example**

Find the volume and surface area of a can of tomato soup where $d$ = 7cm, $h$ = 10cm.

$\frac{1}{2}$ diameter = radius

Volume = $\pi r^2 h = \pi(3.5)^2 \times 10 = 384.8 = 385$cm$^3$

Surface area = $2\pi(3.5 \times 10) + 2\pi(3.5)^2$

$= 219.91 + 76.97$

$= 296.88$

$= 297$cm$^2$ ← Give answer to nearest whole number

## Pyramids

> **KEY POINT**
>
> Volume of a pyramid $= \frac{1}{3} \times$ area of base $\times$ perpendicular height
>
> Surface area of pyramid = sum of area of all faces
>
>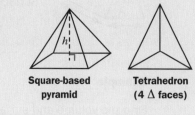
>
> | Square-based pyramid | Tetrahedron (4 △ faces) |

A pyramid usually takes its name from the shape of its base.

## Cones

> **KEY POINT**
>
> Volume of a cone $= \frac{1}{3} \times$ area of base $\times$ perpendicular height
>
>  $V = \frac{1}{3}\pi r^2 h$ where $r$ is the radius of the base
>
> Curved surface area $= \pi r l$ where $l$ is the slant height

## Frustum of a cone

> **KEY POINT**
>
> A frustum of a cone is produced when the top of the cone is sliced parallel to the base.
>
> 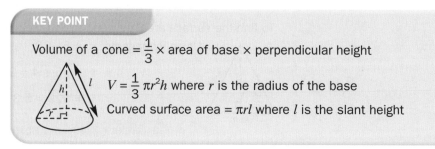 Volume of a frustum
> = volume of whole cone − volume of removed cone
> $= \frac{1}{3}\pi R^2 H - \frac{1}{3}\pi r^2 h$
>
> Curved surface area of frustum
> $= \begin{array}{c} \text{curved surface area} \\ \text{of the whole cone} \end{array} - \begin{array}{c} \text{curved surface area} \\ \text{of removed cone} \end{array}$
> $= \pi R L - \pi r l$ where $l$ is the slant height

You may need to add the areas of the circles at the top and bottom of the frustum. Read the question carefully.

## Spheres

> **KEY POINT**
>
> Volume of sphere $= \frac{4}{3}\pi r^3$ where $r$ is the radius of the sphere
>
> Surface area of sphere $= 4\pi r^2$
>
>
>
> Halve the volume of a sphere to find the volume of a hemisphere.

# Volume and surface area of compound solid shapes

## KEY POINT

To find the volume and surface area of compound solid shapes add together all the components required. It may not be every face. Always check that units are consistent.

## Example

A baby's toy consists of a cone on top of a hemisphere. Calculate its volume and surface area.

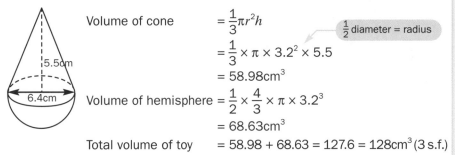

Volume of cone $= \frac{1}{3}\pi r^2 h$

$\frac{1}{2}$ diameter = radius

$= \frac{1}{3} \times \pi \times 3.2^2 \times 5.5$

$= 58.98\text{cm}^3$

Volume of hemisphere $= \frac{1}{2} \times \frac{4}{3} \times \pi \times 3.2^3$

$= 68.63\text{cm}^3$

Total volume of toy $= 58.98 + 68.63 = 127.6 = 128\text{cm}^3 \,(3\text{ s.f.})$

Slant height is needed to find surface area. Using Pythagoras' theorem:

$l^2 = 5.5^2 + 3.2^2 = 30.25 + 10.24 = 40.49$

$l = \sqrt{40.49} = 6.36\text{cm}$

Curved surface area of cone $= \pi r l = \pi \times 3.2 \times \sqrt{40.49}$

$= 63.97\text{cm}^2$

Curved surface area of hemisphere $= \frac{1}{2}(4\pi r^2) = \frac{1}{2} \times 4\pi \times 3.2^2$

$= 64.34\text{cm}^2$

Flat surface of hemisphere is not needed in this calculation.

Total curved surface area of toy $= 63.97 + 64.34 = 128.31\text{cm}^2 = 128\text{cm}^2 \,(3\text{ s.f.})$

## PROGRESS CHECK

1. A decorative box of sweets is made from a cube, sides of 10cm.
   **(a)** What is the total capacity of the box?
   **(b)** What is the area of card needed to make the box?
2. Find the volume and total surface area of a solid rod made from a cylinder of length 16cm and radius 75mm with a hemisphere on each end.
3. What is the capacity of a plant pot formed from an inverted cone, diameter 18cm and height 35cm, with its top, diameter 11cm and height 15cm, sliced off?

3. Capacity = 2494cm³
2. Volume = 4596cm³; surface area = 1461cm²
1. (a) Capacity of box = 1000cm³; (b) Area of card = 600cm²

# 4.7 Effect of enlargement

**LEARNING SUMMARY**

After studying this section, you should be able to understand:

- the effect of enlargement on length, area and volume

## Effect of enlargement on length, area and volume

| | |
|---|---|
| AQA UNITISED | ✓ |
| AQA LINEAR | ✓ |
| EDEXCEL A | ✓ |
| EDEXCEL B | ✓ |
| OCR A | ✓ |
| OCR B | ✓ |
| WJEC UNITISED | ✓ |
| WJEC LINEAR | ✓ |
| CCEA | ✓ |

Length, area and volume can be enlarged by multiplying by a **scale factor**.

- **Length**

  Enlarge the length 2cm by scale factor 3.5

$2 \times 3.5 = 7$

Lengths are in the ratio 2 : 7 or 1 : 3.5

- **Area**

  Enlarge a square, of sides 2.5cm, by scale factor 2 (i.e. multiply each length by 2). What is the effect on the area?

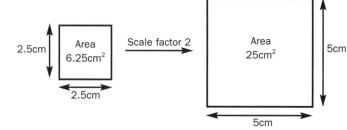

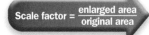
Scale factor = $\dfrac{\text{enlarged area}}{\text{original area}}$

Area enlarges by scale factor $\dfrac{25}{6.25} = 4 = 2^2$

Areas are in the ratio 2 : 2²

**KEY POINT**

If the side of a shape enlarges by scale factor $a$, the area enlarges by scale factor $a^2$.

- **Volume**

  Enlarge a cube, of sides 2.5cm, by scale factor 2. What is the effect on the volume?

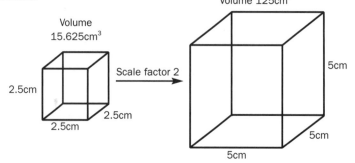

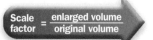
Scale factor = $\dfrac{\text{enlarged volume}}{\text{original volume}}$

Volume enlarges by scale factor $\dfrac{125}{15.625} = 8 = 2^3$

Volumes are in the ratio 2 : 2³

**KEY POINT**

If the side of a shape enlarges by scale factor $a$, the volume enlarges by scale factor $a^3$.

# Sample GCSE questions

**1** The side of a building is to be painted.

    8m    5.7m    10m

  **(a)** Work out the area of the painted surface. **(2)**

  **(b)** A can of paint, containing 1 litre, will cover 12 square metres of wall. How many cans will be needed for one coat of paint? **(1)**

  **(c)** The far side of the building has the same dimensions, but has two rectangular windows with dimensions 1.5m × 1m, two rectangular windows with dimensions 2m × 1.4m, and a door with dimensions 0.75m × 2m. How much paint is needed for this side? **(4)**

  (a) Divide the side into a rectangle and a triangle.

**Height of triangle**
**= 8 − 5.7 = 2.3**

Area of side = $(10 × 5.7) + (\frac{1}{2} × 10 × 2.3)$

           $= 57 + 11.5 = 68.5m^2$

  (b) Number of cans = 68.5 ÷ 12 = 5.7

**You need to consider if**
**your answer is sensible**

You need to buy 6 cans as you cannot buy part of a can of paint.

  (c) Area of windows = 2(1.5 × 1) + 2(2 × 1.4)

                 $= 3 + 5.6 = 8.6m^2$

**The windows and door**
**do not need painting**

Area of door     $= 2 × 0.75 = 1.5m^2$

Area remaining   $= 68.5 − (1.5 + 8.6) = 58.4m^2$

Number of cans = 58.4 ÷ 12 = 4.87

You need to buy 5 cans to paint this side of the building.

**2** A map has a scale of 1 : 1 000 000

  **(a)** Fill in this table of distances between towns G, D and J. Give all answers to 1 d.p. **(3)**

| Towns | Map distance (cm) | Actual distance (km) |
|-------|-------------------|----------------------|
| G → D | 5cm | 50km |
| G → J | 2.4cm | 23.7km |
| J → D | 7.5cm | 75km |

  **(b)** A car drives from G to D at 35mph, stops for half an hour, then continues to J at 30mph. How long did it take to travel from G to J? **(2)**

  **(c)** What was the average speed of the whole journey? **(2)**

  **(d)** The car does 40 miles to the gallon. Petrol costs 108.9p per litre. How much will the petrol cost for the journey? **(2)**

  (a) See answers in table

  (b) mph needed so change distances from km to miles using 5 miles ≈ 8km

**Time = $\dfrac{\text{distance}}{\text{speed}}$**

Time to go from G → D = 31.3 miles ÷ 35mph = 0.89 hours

                        = 0.89 × 60 = 54mins

Time to go from D → J = 46.9 miles ÷ 30mph = 1.56 hours

**Use factor of 60 to**
**change units of time**

                        = 1.56 × 60 = 94mins

Total journey time from G to J = 54mins + 30mins + 94mins

                        = 178mins = 2hrs 58mins

  (c) Average speed of the whole journey = total distance ÷ total time taken

                        $= \dfrac{78.2 \text{ miles}}{178\text{mins}} = \dfrac{78.2 \text{ miles}}{2.97\text{hrs}}$

                        = 26.3mph

**1 gallon ≈ 4.5 litres**

  (d) Number of gallons = 78.2 ÷ 40

Number of litres   $= \dfrac{78.2}{40} × 4.5 = 8.8$

**Divide by 100 to change**
**pence to £. Always**
**give £ to 2 d.p.**

Cost of petrol     = 8.8 × 108.9 = 958p ∴ cost of petrol = £9.58

# Exam practice questions

**1** Fiona is going to paint her living room. It has the dimensions 5.8m by 6.7m by 2.5m. Each can of paint covers 15m$^2$ and costs £11.60.

    **(a)** What is the area of walls and ceiling to be painted, ignoring doors and windows? **(3)**

    **(b)** How many cans of paint will Fiona need to buy? **(2)**

    **(c)** How much will Fiona have to spend? **(1)**

**2** **(a)** A beaker is a cylinder shape. The diameter of the base is 5.5cm and the height is 10cm.

       A jug holds 2.5 litres. How many times can it fill the beaker? **(2)**

    **(b)** A can of sweetcorn has the same dimensions.

       The label covers the curved surface. What is the area of the label to 1 d.p.? **(2)**

    **(c)** What is the total surface area of the can to 3 s.f.? **(3)**

**3** P is a point due east of a harbour H. Q is a point on the coast 10km due south of H.
The distance from P to Q is 14km.
Draw a sketch diagram for this question.
Calculate PH and the bearing of Q from P. **(5)**

**4** A photograph is taken with a digital camera. Its size on the computer screen is 21cm by 15cm.
Calculate the scale factor of enlargement if photographs of the following sizes are required for printing.

    **(a)** 10.5cm by 7.5cm **(1)**

    **(b)** 15.75cm by 11.25cm **(1)**

    **(c)** 26.25cm by 18.75cm **(1)**

    **(d)** 31.5cm by 22.5cm **(1)**

    **(e)** 42cm by 30cm **(1)**

**5** Three spheres of radius 4cm fit exactly in a cylindrical tube. Calculate the volume inside the tube not filled by the spheres. Give your answer to 3 s.f.

    **(5)**

**6** A diagram of a toy sailing boat is in the shape of a quadrant of a circle.

    **(a)** Calculate the perimeter and area of the whole shape. **(5)**

    **(b)** Calculate the area of the boat only. **(2)**

# 5 Statistics

**The following topics are covered in this chapter:**

- Collecting and organising data
- Presenting and interpreting data
- Averages

# 5.1 Collecting and organising data

**LEARNING SUMMARY**

After studying this section, you should be able to understand:

- the data handling cycle
- types of data
- sampling
- collection and organisation

## The data handling cycle

| | |
|---|---|
| AQA UNITISED | ✓ |
| AQA LINEAR | ✓ |
| EDEXCEL A | ✓ |
| EDEXCEL B | ✓ |
| OCR A | ✓ |
| OCR B | ✓ |
| WJEC UNITISED | ✓ |
| WJEC LINEAR | ✓ |
| CCEA | ✓ |

The **data handling cycle** is the process used to solve a statistical problem.

1. Specify the problem / hypothesis
2. Plan
3. Collect the data
4. Process and present the data
5. Interpret and analyse the data
6. Refine problem / hypothesis and / or make conclusion

## Types of data

| | |
|---|---|
| AQA UNITISED | ✓ |
| AQA LINEAR | ✓ |
| EDEXCEL A | ✓ |
| EDEXCEL B | ✓ |
| OCR A | ✓ |
| OCR B | ✓ |
| WJEC UNITISED | ✓ |
| WJEC LINEAR | ✓ |
| CCEA | ✓ |

Items of information are referred to as **data**.

- **Raw data** is unprocessed data which has not been organised, for example the results of an experiment before they have been organised.
- **Discrete data** is separate or distinct items or groups of data that can only have a certain value, for example shoe sizes or colours of cars.
- **Continuous data** is data that has been arranged into groups with no gaps that can have any value, for example measurements of heights of plants.
- **Grouped data** is information that has been organised into groups, for example age groups of a population.
- **Primary data** is original information that is collected from questionnaires, surveys or experiments, for example results obtained by tossing a coin.
- **Secondary data** is information that has been taken from printed books and tables or the Internet, for example league tables of schools.
- **Qualitative data** is descriptive data, for example types of vehicles in a car park.

- **Quantitative data** is data that gives measurement. This can be either in discrete or continuous form, for example the number of pupils in a class is discrete quantitative data and the height of pupils in a class is continuous quantitative data.

# Sampling

| | |
|---|---|
| AQA UNITISED | ✓ |
| AQA LINEAR | ✓ |
| EDEXCEL A | ✓ |
| EDEXCEL B | ✓ |
| OCR A | ✗ |
| OCR B | ✓ |
| WJEC UNITISED | ✗ |
| WJEC LINEAR | ✗ |
| CCEA | ✓ |

> Although population usually refers to the number of people living in a place, it is also used for a large group of items under investigation.

When conducting a survey, it is impossible to deal with large numbers. It is more appropriate to use a **sample** instead. The sample must be large enough to significantly represent the whole group or population.

A sample should be **unbiased**, so must not tend towards particular results. For example, asking opinion about a new product after free samples are given would bias the results.

Sampling can be done in different ways.

## Random sampling

**Random sampling** means having an equal chance of being selected. It is very difficult to get a perfectly random sample, although a computer can generate random numbers.

## Systematic sampling

**Systematic sampling** uses a rule to pick a sample, for example choosing to question every 10th person walking down a street. This may be biased by choosing a particular street or standing in a particular place. The weather or time of day can also change a sample.

## Stratified sampling

**Stratified sampling** takes into account the differences in the composition of a population. This is used to eliminate bias. Samples are taken from each part of the population. They are proportional to the size of the group.

> **KEY POINT**
>
> Stratified sample of a group $= \dfrac{\text{size of group}}{\text{size of population}} \times \text{size of sample}$

For example:
Different travel arrangements of pupils in a school are to be surveyed.
Total number in school = 1100. It is decided to use a sample of 100.

> Remember to calculate the stratified sample to the nearest whole number. You cannot have part of a person!

| Year | Number in year | Stratified sample |
|---|---|---|
| 7 | 190 | $\dfrac{190}{1100} \times 100 = 17$ |
| 8 | 220 | $\dfrac{220}{1100} \times 100 = 20$ |
| 9 | 250 | $\dfrac{250}{1100} \times 100 = 23$ |
| 10 | 240 | $\dfrac{240}{1100} \times 100 = 22$ |
| 11 | 200 | $\dfrac{200}{1100} \times 100 = 18$ |

## Quota sampling

Market researchers tend to use **quota sampling**. They may choose to question a given number of men in a certain age group or a group of people in a particular occupation. Bias can be introduced by the choice of interviewees.

# Collection and organisation

AQA UNITISED ✓
AQA LINEAR ✓
EDEXCEL A ✓
EDEXCEL B ✓
OCR A ✓
OCR B ✓
WJEC UNITISED ✓
WJEC LINEAR ✓
CCEA ✓

**KEY POINT**

Data collection can be done by designing a questionnaire for a survey, observation or experiment. Responses can also be recorded in a data collection sheet or table. The information collected produces a database.

## Questionnaire

A questionnaire is a set of questions designed to find data about a particular topic. The questions must be easy to understand, unbiased and asked in a logical order.

They should produce answers that are easy to analyse.
- Use **closed questions** as they can be answered by a single word or phrase, for example 'How many CDs do you have?'
- Try not to use **open questions** as they allow for many answers, for example, 'What do you think of this CD?'
- Do not use **leading questions** as they put the questioner's opinion into the mind of the interviewee, for example 'Don't you think ten CDs is too many to have of one band?'
- **Response boxes** should be provided for answers where possible, for example yes ❑ no ❑ ; 1 ❑ 2 ❑ 3 ❑

## Observation

If collecting data by observation, write a list of what data you need in order to answer the given question. For example, to answer the question 'How many vehicles pass a certain point during a certain time?' you would need to...

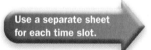
Use a separate sheet for each time slot.

- make a list of all possible vehicles
- divide the total time into smaller slots
- collect data from both sides of the road
- record the vehicles on a data collection sheet.

## Organising data

- A **frequency table** is any table displaying primary or secondary observation data, which is arranged in columns to show the frequency of events.

For example:

Use 'other' in the table for any vehicle not included, such as a caravan.

| Vehicles (10–11am) | Frequency | Vehicles (10–11am) | Frequency |
|---|---|---|---|
| Bike | 3 | Taxi | 2 |
| Motorbike | 2 | Van | 4 |
| Car | 13 | Truck | 3 |
| Bus | 5 | Other | 1 |
| | | **Total** | **33** |

- A **tally chart** has a mark for each observation. This is useful when conducting an experiment. When filling in tally charts, mark I for each vehicle. The 5th mark goes across the previous four marks to make ЖII. It is easier to count in 5s, so mistakes are avoided.

| Vehicle | Tally | Frequency |
|---------|-------|-----------|
| Bike | III | 3 |
| Motorbike | II | 2 |
| Car | ЖII ЖII III | 13 |
| Bus | ЖII | 5 |
| Taxi | II | 2 |
| Van | IIII | 4 |
| Truck | III | 3 |
| Other | I | 1 |
| **Total** | **10–11am** | **33** |

- A **two-way table** illustrates two different variables. This two-way table records whether the vehicles, in the above survey, were used privately or commercially.

| Vehicle | Bike | Motorbike | Car | Bus | Taxi | Van | Truck | Other |
|---------|------|-----------|-----|-----|------|-----|-------|-------|
| **Private** | 3 | 1 | 11 | 0 | 0 | 1 | 0 | 1 |
| **Commercial** | 0 | 1 | 2 | 5 | 2 | 3 | 3 | 0 |

## Class intervals

When collecting continuous data, such as heights or weights, it is convenient to divide it into groups or **class intervals**. For example, this table shows the percentage of women of each age group in part-time work:

> Make sure that the upper class limit is not used as the lower class limit in the next interval. You cannot have 21–25 and 25–30 and so on in this table.

| Age | 21–24 | 25–29 | 30–34 | 35–39 | 40–44 | 45–49 | 50–54 | 55–59 |
|-----|-------|-------|-------|-------|-------|-------|-------|-------|
| % | 20 | 32 | 50 | 52 | 48 | 47 | 52 | 58 |

The class intervals are usually even, but not always.

## Experiment

> **KEY POINT**
>
> A **hypothesis** is a theory that is tested to see if it is true.

To test a hypothesis, you need to set up an experiment. Make sure the experiment produces enough results to produce a significant conclusion.

For example:

Hypothesis: 'Height affects shoe size.'

1. Measure height and shoe size using samples of different groups of people. Make sure the sample is of a sufficient size and reflects the total population being tested.

2. Record results on a data collection sheet.

3. Organise data into a table, such as a tally chart or two-way table.

4. Decide whether the hypothesis is true based on the results.

---

**PROGRESS CHECK**

1. What type of data are the following?

   (a) Long jump lengths recorded at the last Olympic Games.

   (b) Types of sandwiches sold at a café during one week.

   (c) Brands of cereal available at a local supermarket.

2. The following questions are taken from a survey on TV viewing habits. Suggest how they can be improved.

   (a) How old are you?

   (b) What type of programme do you prefer?

   Comedy ❑    Drama ❑    Sport ❑    Politics ❑

   (c) Do you agree that TV programmes are better than they used to be?

3. A health club owner collects data about his 1200 members.

| Age | <30 | 30–40 | 41–50 | >50 |
|-----|-----|-------|-------|-----|
| Male | 205 | 310 | 152 | 53 |
| Female | 124 | 196 | 110 | 50 |

He would like to take a stratified sample of 60 people based on age and gender.

(a) Calculate the number of people from each group that would be used in his sample.

(b) What do you notice about your answers?

(b) Stratified sample adds up to 62 because of rounding up to whole numbers.

| Age | <30 | 30–40 | 41–50 | >50 |
|-----|-----|-------|-------|-----|
| Male | $\frac{205}{1200} \times 60 = 10$ | $\frac{310}{1200} \times 60 = 16$ | $\frac{152}{1200} \times 60 = 8$ | $\frac{53}{1200} \times 60 = 3$ |
| Female | $\frac{124}{1200} \times 60 = 6$ | $\frac{196}{1200} \times 60 = 10$ | $\frac{110}{1200} \times 60 = 6$ | $\frac{50}{1200} \times 60 = 3$ |

3. (a)

Better ❑  Worse ❑  Don't know ❑

you think that TV programmes are better or worse than they used to be?' included (c) Biased and leading question – change it to an unbiased, closed question, e.g. 'Do be should 'Other' (b) etc. ❑, 10–20 ❑, 10 < : ages of groups Give (a) 2. discrete qualitative,

1. (a) Secondary, quantitative, continuous (b) Primary, qualitative, discrete (c) Primary,

# 5.2 Presenting and interpreting data

**After studying this section, you should be able to understand:**

- charts
- diagrams
- graphs

## Charts

AQA UNITISED ✓
AQA LINEAR ✓
EDEXCEL A ✓
EDEXCEL B ✓
OCR A ✓
OCR B ✓
WJEC UNITISED ✓
WJEC LINEAR ✓
CCEA ✓

Once data has been recorded and sorted in a table, it can be presented in the form of a chart.

### Pictogram

A **pictogram** is a chart that uses pictures to represent the numbers of items.

**Example**

The following table gives the number of red, blue, silver and black cars passing a certain point. Present the information as a pictogram.

| Colour of car | Red | Blue | Silver | Black |
|---|---|---|---|---|
| Frequency | 20 | 13 | 10 | 6 |

> You must include a key to your pictures.

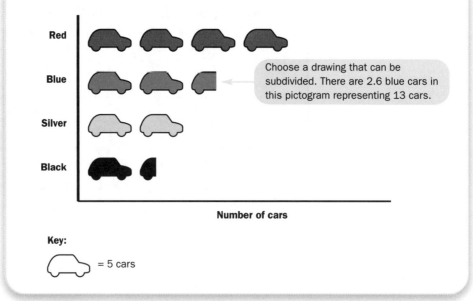

Choose a drawing that can be subdivided. There are 2.6 blue cars in this pictogram representing 13 cars.

Key:

⬤ = 5 cars

### Bar chart

> If bars are used horizontally, the axes are reversed.

A **bar chart** uses bars of equal width to represent frequency. The bars are usually vertical. Frequency is marked on the vertical axis. Items or groups of items are marked on the horizontal axis. Use a bar chart if the number of items is small.

For example, this bar chart shows the number of rainy days over four months.

| Number of rainy days | Jul | Aug | Sept | Oct |
|---|---|---|---|---|
| Frequency | 12.5 | 8 | 4 | 12 |

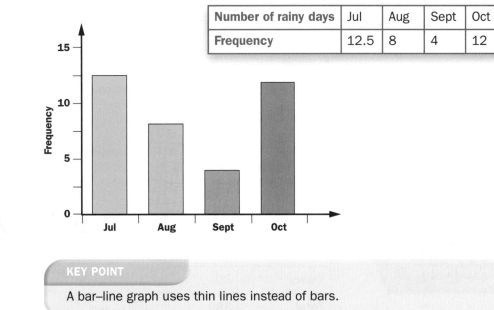

A bar–line graph uses thin lines instead of bars.

Two or more distributions can be compared by using a **dual bar chart** (also known as a comparative or multiple bar chart). For example, this dual bar chart compares across and down clues in a crossword. It shows that the down clues generally have fewer letters in their answer.

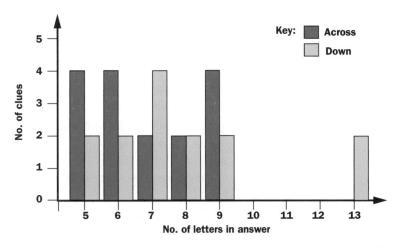

A **composite bar chart** compares two or more distributions on each bar. For example, this composite bar chart compares the activities chosen by students in each school year. The total number of students in each year can be read off the individual bars, e.g. there are 90 students in year 7.

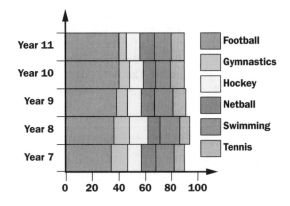

## Pie chart

A **pie chart** is a circular chart that uses sectors in proportion to value.

**Example**

Draw a pie chart to show the way in which 360 pupils travel to the local school.

Sector angle $= \dfrac{\text{value}}{\text{total}} \times 360°$

| Travel method | Frequency |
|---|---|
| Walk | 135 |
| Bus | 120 |
| Train | 75 |
| Cycle | 30 |
| **Total** | **360** |

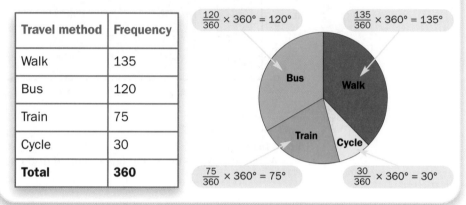

$\dfrac{120}{360} \times 360° = 120°$   $\dfrac{135}{360} \times 360° = 135°$

$\dfrac{75}{360} \times 360° = 75°$   $\dfrac{30}{360} \times 360° = 30°$

> Mark the centre and one radius. Measure the first angle from here.
>
> The sum of the sector angles should equal 360°.
>
> You must label your diagram.

Two or more distributions can be compared by comparing sectors of their pie charts. A pie chart only shows proportions so remember to take the individual populations into account.

## Histogram

A **histogram** is a chart that uses bars to represent continuous grouped data. There are no gaps between the bars in a histogram.

- **Equal class intervals**

  If the class intervals are equal, frequency is proportional to the heights of the bars. The vertical axis shows the frequency. The horizontal axis is divided into equal class intervals.

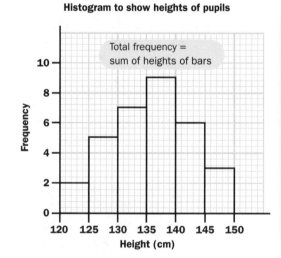

**Histogram to show heights of pupils**

Total frequency = sum of heights of bars

| Height (cm) | Frequency |
|---|---|
| $120 \leqslant h < 125$ | 2 |
| $125 \leqslant h < 130$ | 5 |
| $130 \leqslant h < 135$ | 7 |
| $135 \leqslant h < 140$ | 9 |
| $140 \leqslant h < 145$ | 6 |
| $145 \leqslant h < 150$ | 3 |
| **Total** | **32** |

● **Unequal class intervals**

If the class intervals are unequal, frequency is proportional to the areas of the bars. This is called frequency density. Work out frequency density using:

$$\text{Frequency density} = \frac{\text{frequency}}{\text{class width interval}}$$

Draw the histogram using frequency density on the vertical axis and class width on the horizontal axis.

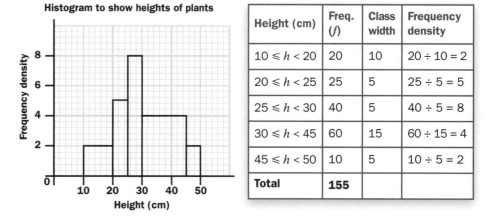

| Height (cm) | Freq. (f) | Class width | Frequency density |
|---|---|---|---|
| $10 \leqslant h < 20$ | 20 | 10 | $20 \div 10 = 2$ |
| $20 \leqslant h < 25$ | 25 | 5 | $25 \div 5 = 5$ |
| $25 \leqslant h < 30$ | 40 | 5 | $40 \div 5 = 8$ |
| $30 \leqslant h < 45$ | 60 | 15 | $60 \div 15 = 4$ |
| $45 \leqslant h < 50$ | 10 | 5 | $10 \div 5 = 2$ |
| **Total** | **155** | | |

If you are provided with a histogram it is possible to find out the frequencies, the modal class and the median of the data.

For example, if you were given the histogram above:

● First draw up a table, filling in the heights, class widths and frequency densities from the histogram.

● Work out frequency (f) as shown below.

| Height (cm) (class interval) | Class width | Frequency density | Frequency (f) (class width × frequency density) |
|---|---|---|---|
| $10 \leqslant h < 20$ | 10 | 2 | $10 \times 2 = 20$ |
| $20 \leqslant h < 25$ | 5 | 5 | $5 \times 5 = 25$ |
| $25 \leqslant h < 30$ | 5 | 8 | $5 \times 8 = 40$ |
| $30 \leqslant h < 45$ | 15 | 4 | $15 \times 4 = 60$ |
| $45 \leqslant h < 50$ | 5 | 2 | $5 \times 2 = 10$ |
| | | **Total** | **155** |

Modal class = $30 \leqslant h < 45$ ← This class has the highest frequency

See pages 160–161 for modal class and median.

The median is the middle value, when placed in ascending order. There are 155 plants, so the middle value is the 78th. This is in the $25 \leqslant h < 30$ class. ← This class goes from the 46th value to the 85th value

## Frequency polygon

To draw a **frequency polygon** you need to plot the frequency at the midpoint of each class interval.

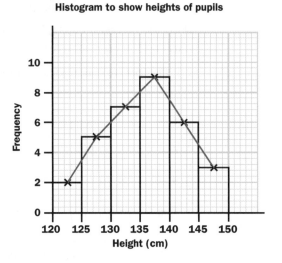

A frequency polygon can be drawn without the histogram.

Distributions can be compared by drawing their frequency polygons on the same axes.

## Diagrams

| | |
|---|---|
| AQA UNITISED | ✓ |
| AQA LINEAR | ✓ |
| EDEXCEL A | ✓ |
| EDEXCEL B | ✓ |
| OCR A | ✓ |
| OCR B | ✓ |
| WJEC UNITISED | ✓ |
| WJEC LINEAR | ✓ |
| CCEA | ✓ |

The distribution of data can be displayed in the form of a diagram.

## Stem and leaf diagram

A stem and leaf diagram is used for displaying data. It is useful for showing the shape of a distribution.

For example:
Here are the marks gained by 12 pupils in a Maths test.

| 20 | 30 | 20 | 15 | 22 | 17 | 16 | 22 | 27 | 8 | 29 | 25 |
|---|---|---|---|---|---|---|---|---|---|---|---|

The marks can be illustrated on a stem and leaf diagram.
- The tens digit of each value forms the 'stem'.
- The units digit of each value forms the 'leaf'.

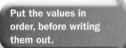

Put the values in order, before writing them out.

```
0 | 8                          ← Mark is 8
1 | 5  6  7                     ← Marks are 15, 16, 17
2 | 0  0  2  2  5  7  9         ← Marks are 20, 20, 22, and so on
3 | 0                           ← Mark is 30
```

From the stem and leaf diagram above, you can identify certain information:
- The modal group, the one with the highest frequency, is the 20–29 group.
- There are 12 results so the median result is mid-way between the 6th and 7th result. The 6th result is 20 and the 7th result is 22 so the median is 21.

See pages 160–161 for modal class and median.

Two distributions may be compared by using a common stem, with leaves that go back-to-back.

Box plot diagrams are sometimes called box and whisker diagrams.

See pages 160–162 for median and quartiles.

# Box plot diagram

Box plot diagrams display shape, range, median and quartiles.

For example:

A maths teacher recorded the times that a Year 7 class spent on their maths homework one night. The table shows the times, to the nearest minute, after they have been arranged in order, from smallest to largest:

| 12 | 16 | 16 | 18 | 18 | 18 | 18 | 19 | 19 |
|----|----|----|----|----|----|----|----|----|
| 19 | 20 | 20 | 21 | 21 | 21 | 21 | 21 | 21 |
| 25 | 26 | 27 | 29 | 29 | 30 | 30 |

The smallest value = 12, the largest value = 30.
The median is the 13th value = 21.
The lower quartile is the median of the data to the left of the actual median:

| 12 | 16 | 16 | 18 | 18 | 18 | 18 | 19 | 19 | 19 | 20 | 20 |
|----|----|----|----|----|----|----|----|----|----|----|----|

Lower quartile = 18

The upper quartile is the median of the data to the right of the actual median:

| 21 | 21 | 21 | 21 | 21 | 25 | 26 | 27 | 29 | 29 | 30 | 30 |
|----|----|----|----|----|----|----|----|----|----|----|----|

Upper quartile = 25.5

So the lower quartile = 18 and the upper quartile = 25.5

You can now draw a box plot for the data. The 'box' stretches from 18, (the lower quartile), to 25.5, (the upper quartile). The median is shown inside the box at 21, and the 'whiskers' stretch from the lower quartile to the smallest value, 12, and from the upper quartile to the highest value, 30.

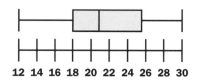

Box plots can be very useful in giving a quick 'picture' of a distribution and for comparing two distributions:

- If the median is in the middle of the box, the distribution of data is symmetrical.
- If the median is nearer the lower quartile, the distribution of data is positively skewed.
- If the median is nearer the upper quartile, the distribution of data is negatively skewed.

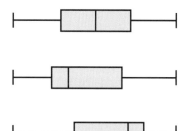

# Graphs

Graphs are another way of presenting data.

## Scatter graph

A **scatter graph** (or scatter diagram) is useful when comparing two sets of variables. The variables are plotted against each other. The relationship between them is called **correlation**.

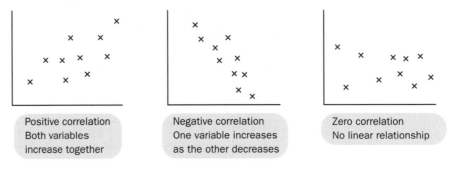

Positive correlation
Both variables
increase together

Negative correlation
One variable increases
as the other decreases

Zero correlation
No linear relationship

> If just one point is outside the general trend, this may be a 'rogue' point, so check it carefully.

If the points are almost in a straight line, there is a strong correlation. Otherwise they may have a moderate or weak correlation.

If there is a strong correlation, a **line of best fit** can be drawn. This shows the trend. It should go through the middle of the set of points. It can be used to predict values beyond those plotted.

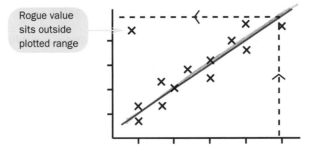

> Rogue value sits outside plotted range

## Time–series graph

**Time–series graphs** show trends over time or seasons. They have time on the horizontal axis.

For example, the changes in a patient's temperature can be shown on a time–series graph.

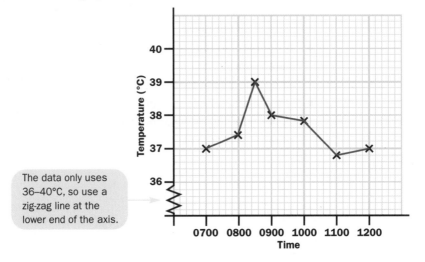

> The data only uses 36–40°C, so use a zig-zag line at the lower end of the axis.

# Cumulative frequency graph

The **cumulative frequency** tells you how often a result was obtained that was less than or equal to a given value in a collection of data. Cumulative frequency is found by adding together the frequencies to give a running total.

For example:
The marks gained in a Maths test were recorded in a cumulative frequency table. The cumulative frequency is found by adding each value to the sum of all the previous values.

| Mark group | Frequency ($f$) | Upper limit of the group | Cumulative frequency |
|---|---|---|---|
| 1–10 | 2 | $\leqslant 10$ | 2 |
| 11–20 | 10 | $\leqslant 20$ | 2 + 10 = 12 |
| 21–30 | 15 | $\leqslant 30$ | 15 + 12 = 27 |
| 31–40 | 20 | $\leqslant 40$ | 20 + 27 = 47 |
| 41–50 | 16 | $\leqslant 50$ | 16 + 47 = 63 |
| 51–60 | 12 | $\leqslant 60$ | 12 + 63 = 75 |

(Total frequency = 75)

> You will usually be given all the values in the question and have to sort them into groups yourself.

A cumulative frequency graph is drawn using the cumulative frequency table. The points are plotted at the upper limit of each group and then connected with a smooth curve.

> Depending on your exam board, a cumulative frequency graph may have points joined using a ruler.

> See page 162 for how to interpret cumulative frequency graphs.

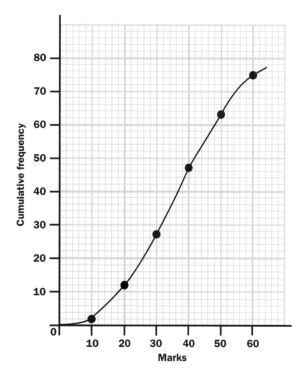

# Misleading graphs

Graphs, charts and diagrams are used to illustrate data. Unfortunately they are sometimes used to distort statistics. Do not take these illustrations at face value.

Beware of **misleading graphs** that use...

- a frequency axis not starting at zero to distort shape
- unequal intervals on the frequency axis
- 3D to change perspective in bar charts, pie charts and pictograms
- unequal sized pictures in pictograms
- axes without labels and graphs without scales
- comparisons that do not start from the same level.

---

**PROGRESS CHECK**

1. Calculate the sector angles for a pie chart, which represents the favourite foods of a group of friends.

| Type of food | Burgers | Chips | Salad | Chocolate | Fish | Pasta |
|---|---|---|---|---|---|---|
| Frequency | 5 | 8 | 3 | 4 | 2 | 8 |

2. Calculate frequency density for this data about heights of trees in a park.

| Height of tree (h) in metres | < 2 | $2 \leqslant h < 3$ | $3 \leqslant h < 5$ | $5 \leqslant h < 7$ | $7 \leqslant h < 8$ | $\geqslant 8$ |
|---|---|---|---|---|---|---|
| Frequency | 1 | 4 | 6 | 5 | 2 | 0 |

3. These scatter graphs illustrate data collected by three different school years.

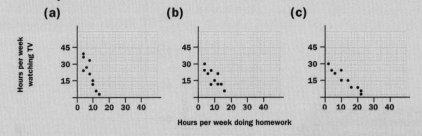

State the correlation for each. What can be said about the relationship between watching TV and doing homework?

3. a) Negative b) Negative c) Negative; The more hours spent watching TV, the less that are spent doing homework.

2.
| Height of tree (h) in metres | < 2 | $2 \leqslant h < 3$ | $3 \leqslant h < 5$ | $5 \leqslant h < 7$ | $7 \leqslant h < 8$ | $\geqslant 8$ |
|---|---|---|---|---|---|---|
| Frequency density | 0.5 | 4 | 3 | 2.5 | 2 | 0 |

1.
| Type of food | Burgers | Chips | Salad | Chocolate | Fish | Pasta |
|---|---|---|---|---|---|---|
| Angle | 60° | 96° | 36° | 48° | 24° | 96° |

# 5.3 Averages

**After studying this section, you should be able to understand:**

- mean, median, mode and range
- quartiles and inter-quartile range
- moving averages

## Mean, median, mode and range

AQA UNITISED ✓
AQA LINEAR ✓
EDEXCEL A ✓
EDEXCEL B ✓
OCR A ✓
OCR B ✓
WJEC UNITISED ✓
WJEC LINEAR ✓
CCEA ✓

The **mean**, **median** and **mode** are all **averages**. Average is the term given to a representative value.

### Mean

> **KEY POINT**
>
> $$\text{Mean} = \frac{\text{total sum of values}}{\text{number of values}}$$

**Examples**

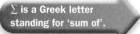
Check that the answer lies between the smallest and largest values.

1. Calculate the mean from the following list of values.
   12.3cm, 14cm, 10.5cm, 11.7cm, 9cm

   $$\text{Mean} = \frac{(12.3 + 14 + 10.5 + 11.7 + 9)}{5} = 11.5\text{cm}$$

   Number of values

2. Calculate the mean from the following frequency table.

| Mins. taken to complete question ($x$) | Number of students ($f$) | $f \times x = fx$ |
|---|---|---|
| 7 | 2 | 14 |
| 9 | 5 | 45 |
| 11 | 4 | 44 |
| 12 | 2 | 24 |
| **Total** | **13** | **127** |

$\Sigma$ is a Greek letter standing for 'sum of'.

Divide total of $fx$ products by total frequency:

$$\text{Mean} = \frac{\Sigma fx}{\Sigma f} = \frac{127}{13} = 9.8\text{mins (2 s.f.)}$$

You cannot calculate the exact value of the mean when the data is grouped or continuous. You can only estimate it using the midpoint of each group.

> **KEY POINT**
>
> $$\text{Estimated mean} = \frac{(\text{midpoint of each group} \times \text{frequency})}{\text{total frequency}}$$

**Example**

A sample of apples is weighed when picked. Calculate an estimated mean from the data in the table below.

| Weight (g) | Frequency ($f$) | Midpoint ($m$) | $fm$ ($f \times m$) |
|---|---|---|---|
| $100 \leqslant w < 110$ | 5 | 105 | 525 |
| $110 \leqslant w < 120$ | 7 | 115 | 805 |
| $120 \leqslant w < 130$ | 12 | 125 | 1500 |
| $130 \leqslant w < 140$ | 8 | 135 | 1080 |
| $140 \leqslant w < 150$ | 4 | 145 | 580 |
| **Total** | **36** | | **4490** |

Estimated mean $= \dfrac{\Sigma fm}{\Sigma f} = \dfrac{4490}{36} = 124.7\text{g}$

## Median

**KEY POINT**

The median is the middle value in an ascending sequence of values.

**Examples**

1. Find the median length of each set of discrete data.
   **(a)** 3cm, 3cm, 1cm, 1.5cm, 2cm, 2cm, 2.5cm

   First, rearrange in ascending order:
   1cm, 1.5cm, 2cm, 2cm, 2.5cm, 3cm, 3cm
   There are seven values so the median is the 4th value.
   The median is 2cm.

   **(b)** 5km, 8km, 4km, 4km, 6km, 6km

   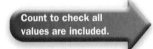

   4km, 4km, 5km, 6km, 6km, 8km ← Rearrange
   There are six values so the median is halfway between the 3rd and 4th values.
   The median is 5.5km.

2. Find the median weight of apples from the data in the table below.

| Weight (g) | Frequency ($f$) |
|---|---|
| $100 \leqslant w < 110$ | 5 |
| $110 \leqslant w < 120$ | 7 |
| $120 \leqslant w < 130$ | 12 |
| $130 \leqslant w < 140$ | 8 |
| $140 \leqslant w < 150$ | 4 |
| **Total** | **36** |

As this is continuous grouped data find the group, or class, that contains the middle value. The middle value is halfway between the 18th and 19th values. Starting with the lowest weight, you will find that the 18th and 19th values are in the $120 \leqslant w < 130$ group.
∴ median group is $120 \leqslant w < 130$

**Count to check all values are included.**

## Mode

> **KEY POINT**
>
> The mode is the most frequent value.

### Examples

**1.** Find the mode of the following discrete data.

0, 3, 2, 5, 5, 2, 3, 1, 2, 0, 2

The most frequent number is 2.
This is the mode.

**2.** Find the mode of weights of apples from the data in the table below.

| Weight (g) | Frequency ($f$) |
|---|---|
| $100 \leqslant w < 110$ | 5 |
| $110 \leqslant w < 120$ | 7 |
| $120 \leqslant w < 130$ | 12 |
| $130 \leqslant w < 140$ | 8 |
| $140 \leqslant w < 150$ | 4 |
| **Total** | **36** |

As this is continuous grouped data, find the group, or class, with the highest frequency.
This is $120 \leqslant w < 130$.
This is called the **modal class**.

## Range

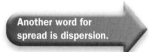

Another word for spread is dispersion.

> **KEY POINT**
>
> The **range** is the spread of data.
> Range = greatest value – least value

### Examples

Find the range of each set of data.
**(a)** 0, 3, 2, 5, 5, 2, 3, 1, 2, 0, 2
    Range = 5 – 0 = 5

**(b)** 12.3cm, 14cm, 10.5cm, 11.7cm, 9cm
    Range = 14cm – 9cm = 5cm

## Using appropriate averages

When choosing which average to use, think about the context of the question:
- The mean gives a typical value. It uses all the values including the extremes.
- The median is useful if there are extreme values.
- The mode gives the most common value.

# Quartiles and inter-quartile range

| | |
|---|---|
| AQA UNITISED | ✓ |
| AQA LINEAR | ✓ |
| EDEXCEL A | ✓ |
| EDEXCEL B | ✓ |
| OCR A | ✓ |
| OCR B | ✓ |
| WJEC UNITISED | ✓ |
| WJEC LINEAR | ✓ |
| CCEA | ✓ |

**KEY POINT**

A **quartile** is one of three values, which divides a frequency distribution into four intervals.

- **Lower quartile** (LQ) is one-quarter of the way through the data.
- **Median** is halfway through the data.
- **Upper quartile** (UQ) is three-quarters of the way through the data.

The **inter-quartile range** (IQ) = upper quartile − lower quartile.

The inter-quartile range measures spread; it is the middle 50% of the distribution of the data.

## Interpreting cumulative frequency graphs

> See page 157 for cumulative frequency graphs.

Once you have drawn a cumulative frequency graph you can compare the distribution of the data.

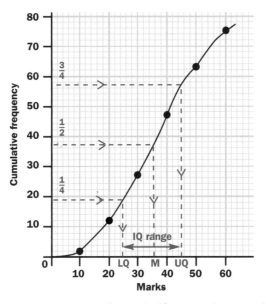

- Mark the median at halfway up the cumulative frequency axis, the upper quartile at three-quarters of the way up and the lower quartile at one-quarter of the way up.
- Draw horizontal dotted lines from these points to meet the curve.
- Draw vertical dotted lines from where they meet the curve to meet the horizontal axis.
- These points give the median, upper quartile and lower quartile of the distribution.

> Use these to compare distributions.

- The inter-quartile range can now be calculated: IQ = UQ − LQ = 45 − 25 = 20

# Moving averages

| | |
|---|---|
| AQA UNITISED | ✗ |
| AQA LINEAR | ✗ |
| EDEXCEL A | ✗ |
| EDEXCEL B | ✗ |
| OCR A | ✗ |
| OCR B | ✓ |
| WJEC UNITISED | ✗ |
| WJEC LINEAR | ✗ |
| CCEA | ✗ |

**Moving averages** are used to look at a trend over time. The average of a block of values is taken.

A four-point moving average uses four data items in each calculation, a three-point moving average uses three and so on.

**Example**

A local garage's sales figures for the past ten years are recorded in the table below.

**(a)** Plot and draw the year and number of cars sold as a time–series graph.

**(b)** Calculate a three-point moving average for the data, plot using the same axes and draw a line of best fit.

| Year | No. of cars sold | Moving averages | | | | | | | |
|---|---|---|---|---|---|---|---|---|---|
| 1 | 70 | 70 | | | | | | | |
| 2 | 65 | 65 | 65 | | | | | | |
| 3 | 72 | 72 | 72 | 72 | | | | | |
| 4 | 70 | | 70 | 70 | 70 | | | | |
| 5 | 56 | | | 56 | 56 | 56 | | | |
| 6 | 58 | | | | 58 | 58 | 58 | | |
| 7 | 65 | | | | | 65 | 65 | 65 | |
| 8 | 58 | | | | | | 58 | 58 | 58 |
| 9 | 60 | | | | | | | 60 | 60 |
| 10 | 68 | | | | | | | | 68 |
| Three-point moving total | | 207 | 207 | 198 | 184 | 179 | 181 | 183 | 186 |
| Three-point moving average | | 69 | 69 | 66 | 61.3 | 59.7 | 60.3 | 61 | 62 |

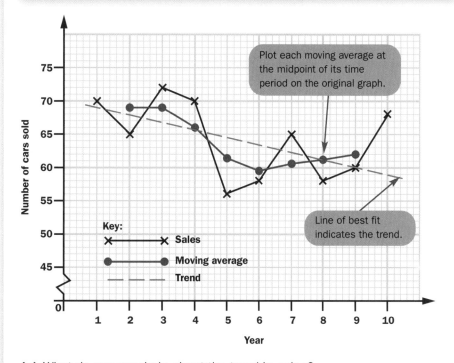

Plot each moving average at the midpoint of its time period on the original graph.

Line of best fit indicates the trend.

**Key:**
✕——✕ Sales
●——● Moving average
— — — Trend

**(c)** What do you conclude about the trend in sales?

Looking at the moving average graph, you can see that the sales trend is downwards, although sales in years 6 to 9 are up. If sales continue to improve, this would alter the moving averages and the trend line.

1 Calculate **(i)** the mean **(ii)** the mode and **(iii)** the median of the following sets of data:

**(a)** Heights: 125cm, 126cm, 126cm, 125cm, 126cm, 127cm, 130cm

**(b)** Weights: 36kg, 37kg, 37kg, 38kg, 39kg, 39kg, 40kg

**(c)** Times: 25s, 26s, 27s, 27s, 28s, 28s, 28s

2 Find **(i)** the modal class and **(ii)** the estimated mean of...

**(a)** heights of flowers

| Height ($h$) cm | < 10 | $10 \leqslant h < 20$ | $20 \leqslant h < 30$ | $30 \leqslant h < 40$ |
|---|---|---|---|---|
| Frequency ($f$) | 20 | 19 | 21 | 4 |

**(b)** test marks

| Marks ($m$) | < 10 | $10 \leqslant m < 15$ | $15 \leqslant m < 20$ | $20 \leqslant m < 25$ | $25 \leqslant m < 30$ |
|---|---|---|---|---|---|
| Freq. ($f$) | 7 | 13 | 8 | 3 | 1 |

3 Work out the cumulative frequencies for this table showing ages of people living on one road. Draw a cumulative frequency graph and find **(i)** the median and **(ii)** the inter-quartile range.

| Age ($a$) | < 15 | $15 \leqslant a < 30$ | $30 \leqslant a < 45$ | $45 \leqslant a < 60$ | $60 \leqslant a < 80$ |
|---|---|---|---|---|---|
| Frequency ($f$) | 14 | 35 | 27 | 16 | 8 |

1. (a) (i) 126.4cm (ii) 126cm (iii) 126cm (b) (i) 38kg (ii) 38kg (iii) 37kg/39kg (c) (i) 27s
(ii) 28s (iii) 27s   2. a) (i) $20 \leqslant h < 30$ (ii) 16.4cm (b) (i) $10 \leqslant m < 15$ (ii) 13.5
3. Cumulative frequencies: 14, 49, 76, 92, 100; (i) median = 31 (ii) IQ range = 24

# Sample GCSE questions

**1** This table shows the sales of drinks from a machine.

(a) What is the mean number of drinks sold of each type per week? **(4)**

(b) Which is the modal drink for the week? Is it the same mode every day? **(1)**

| Drink | Mon | Tue | Wed | Thur | Fri |
|---|---|---|---|---|---|
| Tea | 85 | 90 | 102 | 88 | 82 |
| Coffee | 95 | 92 | 98 | 90 | 88 |
| Chocolate | 30 | 35 | 34 | 36 | 35 |
| Orange | 42 | 42 | 42 | 42 | 42 |
| Lemon | 36 | 40 | 40 | 38 | 35 |
| Milk | 24 | 23 | 23 | 24 | 22 |

(c) Draw a comparative bar chart to illustrate the sale of hot drinks for the week. **(4)**

(d) Compare and comment on these results for the week. **(1)**

> On the exam paper, it may be simpler to add a column to the table for totals

(a)

| Drink | Tea | Coffee | Chocolate | Orange | Lemon | Milk |
|---|---|---|---|---|---|---|
| Total per week | 447 | 463 | 170 | 210 | 189 | 116 |
| Total ÷ 5 = mean | 89.4 | 92.6 | 34 | 42 | 37.8 | 23.2 |

> The mode has the highest total

(b) Modal drink for the week is coffee.

It is not the mode every day. Tea is the mode on Wednesdays.

> Remember to put in a key

(c)

(d) Coffee and tea are the most popular drinks each day (people like a hot drink; it is what they are used to). Milk is the least popular (milk must be kept cool so must be drunk straight away). The same number of orange drinks are sold every day (perhaps it is the favourite of 42 people).

> Give sensible reasons to gain marks

**2** (a) Draw a pie chart to illustrate the number of pets owned by the households on a particular street. Show all your working. **(4)**

| Pet | Bird | Cat | Dog | Fish | Gerbil | Hamster |
|---|---|---|---|---|---|---|
| Frequency | 2 | 8 | 7 | 15 | 5 | 3 |

(b) Ayesha says fish are the most popular because they are pretty to look at. Is she right? Can you give any other reasons for their popularity? **(1)**

> The quickest way to work out the angles for a pie chart is to find the angle for one item

(a) Total number of pets = 40

Angle for one pet = 360° ÷ 40 = 9°

> Remember that the sum of the angles = 360°

> Take care to label the pie chart sectors

9 × 15 = 135°
9 × 5 = 45°
9 × 3 = 27°
9 × 2 = 18°
9 × 8 = 72°
9 × 7 = 63°

Fish, Gerbil, Hamster, Bird, Cat, Dog

(b) She may be right, but they are also easy to look after, they do not need to be taken out for a walk, etc.

# Exam practice questions

**1** Fiona wants to find out how much people spend on downloading music. She needs to design a questionnaire.

**(a)** Write down three questions she might use.

**(b)** She is going to ask all the people in her year. Is this a suitable sample? Give reasons for your answer.  **(3)**

**2** A science test was given to all the students in a year. The progress of both boys and girls was shown by marks being recorded in a back-to-back stem and leaf diagram.

Compare these marks. Give reasons for your conclusions.  **(3)**

| Marks (girls) | Cell | Marks (boys) |
|---|---|---|
| 7 | 1 | 8 |
| 9 3 1 0 | 2 | 8 9 9 |
| 9 8 5 5 3 | 3 | 0 2 5 7 |
| 9 5 4 4 1 0 | 4 | 1 2 3 5 |
| 9 8 7 1 0 0 | 5 | 0 2 3 3 5 9 |
| 7 4 1 0 | 6 | 0 3 5 6 |

**3** The speed of 100 cars, driving on a stretch of motorway, is checked by a camera.

**(a)** Calculate an estimate of the mean speed of these cars using class midpoints. Remembering that the motorway speed limit is 70mph, comment on your answer.  **(4)**

**(b)** Draw a histogram illustrating the results.  **(3)**

| Speed ($s$ mph) | Frequency ($f$) | Midpoint ($m$) |
|---|---|---|
| $40 \leqslant s < 50$ | 4 | |
| $50 \leqslant s < 60$ | 25 | |
| $60 \leqslant s < 70$ | 33 | |
| $70 \leqslant s < 80$ | 35 | |
| $80 \leqslant s < 90$ | 3 | |

**4** The pie charts opposite show the change in population age over 25 years in one town. What changes do you notice?

Give a possible reason for each change.  **(3)**

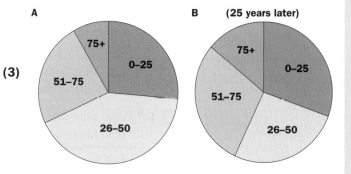

**5** Phil and Dave run a football fan magazine, which is bought on subscription. This is a record of their subscribers over a two-year period.

**(a)** Draw a time–series graph from these end of quarter results.  **(5)**

**(b)** Draw a trend line.  **(1)**

**(c)** Estimate how many subscriptions will be bought in the next three months.  **(1)**

| Year | Months | Frequency |
|---|---|---|
| First year | Jan–Mar | 150 |
| | Apr–Jun | 80 |
| | Jul–Sep | 100 |
| | Oct–Dec | 186 |
| Second year | Jan–Mar | 200 |
| | Apr–Jun | 120 |
| | Jul–Sep | 120 |
| | Oct–Dec | 275 |

# 6 Probability

The following topics are covered in this chapter:

- **Probability**
- **Theoretical probability and relative frequency**
- **Probability rules**
- **Tree diagrams**

# 6.1 Probability

**LEARNING SUMMARY**

After studying this section, you should be able to understand:

- probability vocabulary
- probability scale

## Probability vocabulary

| | |
|---|---|
| AQA UNITISED | ✓ |
| AQA LINEAR | ✓ |
| EDEXCEL A | ✓ |
| EDEXCEL B | ✓ |
| OCR A | ✓ |
| OCR B | ✓ |
| WJEC UNITISED | ✓ |
| WJEC LINEAR | ✓ |
| CCEA | ✓ |

The **probability** of an event occurring is the chance that it may happen. This can be given as a fraction, decimal or percentage. The result of an event is an **outcome**.

> **KEY POINT**
>
> Probability (P) = $\dfrac{\text{number of times an event can happen}}{\text{total number of possible outcomes}}$

**Example**

A dice is thrown. There are six possible outcomes: 1, 2, 3, 4, 5 or 6.

**(a)** What is the probability of throwing a 6?

P(throwing a 6) = $\dfrac{1}{6}$

*If you are giving an answer in fraction form, give it in its lowest terms.*

**(b)** What is the probability of throwing an even number?

P(throwing an even number) = $\dfrac{3}{6} = \dfrac{1}{2}$

*The phrase 'mutually exclusive' may not be used in the exam.*

**Mutually exclusive events** are events that cannot happen at the same time. For example, if a dice is thrown, the events 'obtaining a 1' and 'obtaining an even number' are mutually exclusive as they cannot happen at the same time.

Here is some **vocabulary** that you are likely to come across in probability questions:

- **Certain**: P(certain event) = 1

  For example, P(Tuesday follows Monday) = 1

- **Likely**: P(likely event) is between 0.5 and 1

  For example, P(goal will be scored in Saturday's match) = 0.75

- **Evens**: P(even chance event) = $\frac{1}{2}$ or 0.5 or 50%

  For example, P(getting heads when tossing a coin) = 0.5

- **Unlikely**: P(unlikely event) is between 0 and 0.5

  For example, P(snow will fall in November) = 0.2

- **Impossible**: P(impossible event) = 0

  For example, P(scoring 15 when throwing two dice) = 0

It is also useful to learn the following facts:

- A pack of cards consists of 52 cards. There are two red suits (hearts and diamonds) and two black suits (spades and clubs). Each suit has 13 cards (numbers 2–10, Jack, Queen, King and Ace). There may be two Jokers in the pack, but these are rarely used in questions.

- A fair dice has a $\frac{1}{6}$ chance of showing numbers 1–6 when thrown. A fair coin has an even chance of showing heads or tails when tossed. Biased or weighted dice and coins will not produce these outcomes.

# Probability scale

| | |
|---|---|
| AQA UNITISED | ✓ |
| AQA LINEAR | ✓ |
| EDEXCEL A | ✓ |
| EDEXCEL B | ✓ |
| OCR A | ✓ |
| OCR B | ✓ |
| WJEC UNITISED | ✓ |
| WJEC LINEAR | ✓ |
| CCEA | ✓ |

A range of probabilities can be shown on a **probability scale**.

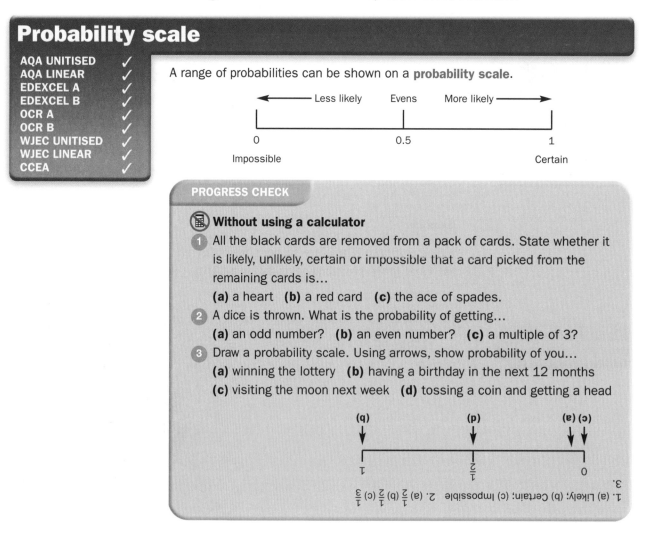

**PROGRESS CHECK**

🔢 **Without using a calculator**

1.  All the black cards are removed from a pack of cards. State whether it is likely, unlikely, certain or impossible that a card picked from the remaining cards is...

    **(a)** a heart   **(b)** a red card   **(c)** the ace of spades.

2.  A dice is thrown. What is the probability of getting...

    **(a)** an odd number?   **(b)** an even number?   **(c)** a multiple of 3?

3.  Draw a probability scale. Using arrows, show probability of you...

    **(a)** winning the lottery   **(b)** having a birthday in the next 12 months

    **(c)** visiting the moon next week   **(d)** tossing a coin and getting a head

3.

1. (a) Likely; (b) Certain; (c) Impossible   2. (a) $\frac{1}{2}$ (b) $\frac{1}{2}$ (c) $\frac{1}{3}$

# 6.2 Theoretical probability and relative frequency

**LEARNING SUMMARY**

**After studying this section, you should be able to understand:**

- estimating probability
- relative frequency graphs
- outcomes of single events
- outcomes of two successive events
- sample space diagrams
- probability of an event not happening

## Estimating probability

**AQA UNITISED** ✓
**AQA LINEAR** ✓
**EDEXCEL A** ✓
**EDEXCEL B** ✓
**OCR A** ✓
**OCR B** ✓
**WJEC UNITISED** ✓
**WJEC LINEAR** ✓
**CCEA** ✓

Probability can be predicted, or estimated, by using experimental data.

- **Theoretical probability** predicts the likelihood of an event occurring if all outcomes are equally likely, i.e. it is what we expect will happen.

> **KEY POINT**
>
> $$\text{Theoretical probability} = \frac{\text{number of times an event can happen}}{\text{total number of possible outcomes}}$$

For example, the probability of throwing a 6 with a fair dice is predicted to be $\frac{1}{6}$.

- **Relative frequency** allows you to estimate how many times an event may occur, during a number of trials. Experimental data can be used to find relative frequency.

> **KEY POINT**
>
> $$\text{Relative frequency} = \frac{\text{number of successful trials}}{\text{total number of trials}}$$

The more times an experiment is repeated, the closer the outcome will be to the theoretical probability. For example, if you throw a fair dice 12 times, you would expect each number to appear twice, because the theoretical probability equals $\frac{1}{6}$ each time. However, it is unlikely that this will happen with only 12 throws. If you were to continue the experiment for 1200 throws though, it should even out so that each number appears 200 times.

If a dice is thrown 600 times and the number 6 appears 200 times, you have to consider if the dice is fair. You would normally expect $\frac{1}{6} \times 600 = 100$ times.

**Example**

Sam drops a drawing pin on the floor 10 times. He notes that it falls on its base 4 times and on its side 6 times.

**(a)** Estimate the number of times the drawing pin will fall on its base and the number of times it will fall on its side if Sam drops the pin 100 times in total.

Number of times pin falls on base = 4 when dropped 10 times

$$\therefore \text{ relative frequency} = \frac{4}{10} = 0.4$$

$$0.4 \times 100 = 40 \text{ when dropped 100 times}$$

Number of times pin falls on side = 6 when dropped 10 times

$$\therefore \text{ relative frequency} = \frac{6}{10} = 0.6$$

$$0.6 \times 100 = 60 \text{ when dropped 100 times}$$

**(b)** Emma drops the same pin 300 times. She works out that the relative frequency of the pin falling on its side is 0.67
How many times did the pin fall on its side?

Use the formula for relative frequency:

Number of times pin falls on side = $0.67 \times 300$ ◁ Relative frequency × total number of trials

$$= 201 \text{ times}$$

# Relative frequency graphs

| | |
|---|---|
| AQA UNITISED | ✓ |
| AQA LINEAR | ✓ |
| EDEXCEL A | ✗ |
| EDEXCEL B | ✗ |
| OCR A | ✗ |
| OCR B | ✗ |
| WJEC UNITISED | ✓ |
| WJEC LINEAR | ✓ |
| CCEA | ✗ |

**Relative frequency graphs** are a useful way of comparing experimental and theoretical data.

For example:

A coin is tossed 150 times. The number of heads after each ten trials is recorded in a table and the relative frequency is plotted on a graph.

> The relative frequency is
> $\dfrac{\text{Total number of heads}}{\text{Total number of trials}}$

| Number of trials ($n$) | Number of Heads ($h$) | Relative frequency ($h \div n$) |
|---|---|---|
| 1–10 | 3 | $\dfrac{3}{10} = 0.3$ |
| 11–20 | 6 | $\dfrac{9}{20} = 0.45$ |
| 21–30 | 7 | $\dfrac{16}{30} = 0.53$ |
| ↓ | ↓ | ↓ |
| 141–150 | 4 | $\dfrac{74}{150} = 0.49$ |

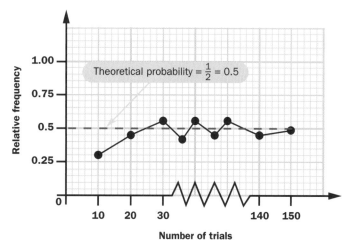

If you do not believe this, try it for yourself! You should find that, as the number of trials increases, the plotted line tends to get closer and closer to the theoretical line.

# Outcomes of single events

AQA UNITISED ✓
AQA LINEAR ✓
EDEXCEL A ✓
EDEXCEL B ✓
OCR A ✓
OCR B ✓
WJEC UNITISED ✓
WJEC LINEAR ✓
CCEA ✓

You may have to find the probability of single events.

### Example

A bag contains 4 red balls, 5 blue balls, 3 green balls and 8 white balls. A ball is picked at random. Find the probability of picking...

**(a)** a green ball

$P(green) = \dfrac{3}{20}$ ← Number of green balls / Total number of balls

**(b)** a red ball

$P(red) = \dfrac{4}{20} = \dfrac{1}{5}$ ← Give in lowest terms. $\dfrac{1}{5}$ can also be given as 0.2 or 20%

**(c)** a black ball

$P(black) = 0$ ← Impossible as there are no black balls

# Outcomes of two successive events

AQA UNITISED ✓
AQA LINEAR ✓
EDEXCEL A ✓
EDEXCEL B ✓
OCR A ✓
OCR B ✓
WJEC UNITISED ✓
WJEC LINEAR ✓
CCEA ✓

## Dependent events

One event may be dependent on the outcome of a previous event. This is called **conditional probability**.

### Example

A bag contains 2 red counters, 2 white counters and 2 blue counters. A counter is picked at random.

**(a)** What is the probability of picking a red counter?

$P(red) = \dfrac{2}{6} = \dfrac{1}{3}$

**(b)** What is the probability of picking a blue counter if the red counter is not replaced?

$P(blue) = \dfrac{2}{5}$ ← There are now only 5 counters as one has been removed

## Independent events

Two events are independent if the outcome of one event has no effect on the outcome of the other event.

### Example

Minnie tosses two coins, one at a time.

**(a)** What is the probability of Minnie tossing a head with the first coin?

$P(head) = \frac{1}{2}$

**(b)** What is the probability of Minnie tossing a head with the second coin?

$P(head) = \frac{1}{2}$

## Sample space diagrams

| AQA UNITISED | ✓ |
| AQA LINEAR | ✓ |
| EDEXCEL A | ✓ |
| EDEXCEL B | ✓ |
| OCR A | ✓ |
| OCR B | ✗ |
| WJEC UNITISED | ✗ |
| WJEC LINEAR | ✗ |
| CCEA | ✗ |

It is often necessary to list all the possible outcomes of two events. This should be done logically otherwise some outcomes may be missed.

All outcomes can be displayed in a **sample space diagram**.

### Example

**(a)** List all the outcomes of throwing two dice.

| | | First dice | | | | | |
| --- | --- | --- | --- | --- | --- | --- | --- |
| | | 1 | 2 | 3 | 4 | 5 | 6 |
| Second dice | 1 | 1/1 | 1/2 | 1/3 | 1/4 | 1/5 | 1/6 |
| | 2 | 2/1 | 2/2 | 2/3 | 2/4 | 2/5 | 2/6 |
| | 3 | 3/1 | 3/2 | 3/3 | 3/4 | 3/5 | 3/6 |
| | 4 | 4/1 | 4/2 | 4/3 | 4/4 | 4/5 | 4/6 |
| | 5 | 5/1 | 5/2 | 5/3 | 5/4 | 5/5 | 5/6 |
| | 6 | 6/1 | 6/2 | 6/3 | 6/4 | 6/5 | 6/6 |

The possible outcomes of each dice are listed in row 1 and column 1. Each cell gives the two numbers on the dice.

**(b)** Find the probability of both dice showing the same number.

Total number of outcomes = 6 × 6 = 36

$P(same number) = \frac{6}{36} = \frac{1}{6}$ ← There are 6 cells with the same two numbers

## Probability of an event not happening

| AQA UNITISED | ✓ |
| AQA LINEAR | ✓ |
| EDEXCEL A | ✓ |
| EDEXCEL B | ✓ |
| OCR A | ✓ |
| OCR B | ✓ |
| WJEC UNITISED | ✓ |
| WJEC LINEAR | ✓ |
| CCEA | ✓ |

**KEY POINT**

The sum of probabilities equals 1 as the total probability is certain. This means that:

P(event will happen) = 1 – P(event will not happen)

or

P(event will not happen) = 1 – P(event will happen)

**Example**

The probability that it will rain on Monday is $\frac{5}{20}$. What is the probability that it will not rain on Monday?

P(will not rain on Monday) = 1 – P(will rain on Monday)

$$= 1 - \frac{5}{20}$$

$$= 1 - \frac{1}{4}$$

$$= \frac{3}{4}$$

**PROGRESS CHECK**

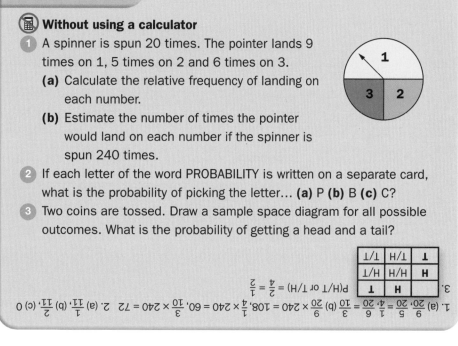

**Without using a calculator**

1. A spinner is spun 20 times. The pointer lands 9 times on 1, 5 times on 2 and 6 times on 3.
   (a) Calculate the relative frequency of landing on each number.
   (b) Estimate the number of times the pointer would land on each number if the spinner is spun 240 times.

2. If each letter of the word PROBABILITY is written on a separate card, what is the probability of picking the letter... (a) P (b) B (c) C?

3. Two coins are tossed. Draw a sample space diagram for all possible outcomes. What is the probability of getting a head and a tail?

3.

|   | H | T |
|---|---|---|
| **H** | H/H | H/T |
| **T** | T/H | T/T |

$P(H/T \text{ or } T/H) = \frac{2}{4} = \frac{1}{2}$

1. (a) $\frac{9}{20}, \frac{5}{20} = \frac{1}{4}, \frac{6}{20} = \frac{3}{10}$ (b) $\frac{9}{20} \times 240 = 108, \frac{1}{4} \times 240 = 60, \frac{3}{10} \times 240 = 72$  2. (a) $\frac{1}{11}$ (b) $\frac{2}{11}$ (c) 0

# 6.3 Probability rules

| **LEARNING SUMMARY** | **After studying this section, you should be able to understand:** |
|---|---|
| | • probability rules |

## Probability rules

| AQA UNITISED | ✓ |
|---|---|
| AQA LINEAR | ✓ |
| EDEXCEL A | ✓ |
| EDEXCEL B | ✓ |
| OCR A | ✓ |
| OCR B | ✓ |
| WJEC UNITISED | ✓ |
| WJEC LINEAR | ✓ |
| CCEA | ✓ |

### Independent events

The outcome of independent events can be worked out using a probability rule.

**KEY POINT**

If A and B are two independent events, then the probability of A and B occurring can be calculated using:

P(A and B) = P(A) × P(B)

If there are more events, they can be included as in:

P(A and B and C) = P(A) × P(B) × P(C)

**Example**

A dice is thrown and a card is picked from a pack.
What is the probability that a 6 is thrown and the Queen of Hearts is picked?

These events are independent of one another.

$P(6) = \frac{1}{6}$    $P(\text{Queen of Hearts}) = \frac{1}{52}$

$P(6 \text{ and Queen of Hearts}) = \frac{1}{6} \times \frac{1}{52} = \frac{1}{312}$

## Mutually exclusive events

Mutually exclusive events can be worked out using a probability rule.

**KEY POINT**

If A and B are mutually exclusive events, then the probability of A or B occurring can be calculated using:
$P(A \text{ or } B) = P(A) + P(B)$

**Example**

A dice is thrown. What is the probability that a 3 or 5 is thrown?

$P(3) = \frac{1}{6}$    $P(5) = \frac{1}{6}$

$P(3 \text{ or } 5) = \frac{1}{6} + \frac{1}{6} = \frac{2}{6} = \frac{1}{3}$

> This is the opposite of what you would expect. 'And' does not mean 'add'.

The two probability rules on pages 173–174 are also known as the **AND / OR rules**.
- If 'and' is used in the question P(A) *and* P(B), then multiply probabilities.
- If 'or' is used in the question P(A) *or* P(B), then add probabilities.

**PROGRESS CHECK**

**Without using a calculator**

1. A dice has letters A–F on its faces. It is thrown and a coin is tossed. What is the probability of getting...
   **(a)** a vowel and a tail? **(b)** an E or F together with a head or tail?
2. A game is played using only the picture cards (King, Queen, Jack) taken from a pack of playing cards. What is the probability of choosing...
   **(a)** a Jack
   **(b)** a red Queen
   **(c)** a black King or a black Queen
   **(d)** a red Jack and a red Queen
   **(e)** an Ace

1. (a) $\frac{1}{6}$ (b) $\frac{1}{3}$ 2. (a) $\frac{4}{12} = \frac{1}{3}$ (b) $\frac{2}{12} = \frac{1}{6}$ (c) $\frac{1}{3}$ (d) $\frac{1}{3}$ (e) 0

# 6.4 Tree diagrams

## Drawing and using tree diagrams

| AQA A | ✓ |
|---|---|
| AQA B | ✓ |
| EDEXCEL A | ✓ |
| EDEXCEL B | ✓ |
| OCR A | ✓ |
| OCR B | ✓ |
| WJEC | ✓ |
| WJEC LINEAR | ✓ |
| CCEA | ✓ |

A **tree diagram** can be used to show all the possible outcomes of an event. Tree diagrams can illustrate dependent, independent and mutually exclusive events.

Each branch on a tree diagram gives a possible outcome for an event. The probability is written on the branch.

### Independent and mutually exclusive events

**Example**

A coin is tossed three times. A tree diagram represents these events.
**(a)** Find the probability of getting three tails
**(b)** Find the probability of getting at least one head and one tail.

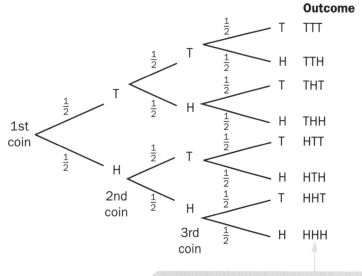

Follow a path along the branches to reach each outcome

**(a)** Multiply probabilities for outcome of three throws.
(tail *and* tail *and* tail → use AND rule)

$$P(\text{three tails}) = \frac{1}{2} \times \frac{1}{2} \times \frac{1}{2} = \frac{1}{8}$$

**(b)** There are six outcomes giving at least one head and one tail.
(TTH *or* THT *or* THH *or* HTT *or* HTH *or* HHT → use OR rule)

$$P(\text{at least one head and one tail}) = \frac{1}{8} + \frac{1}{8} + \frac{1}{8} + \frac{1}{8} + \frac{1}{8} + \frac{1}{8} = \frac{6}{8} = \frac{3}{4}$$

# Dependent events (conditional probability)

**Example**

A bag contains 10 green marbles and 5 yellow marbles.

A marble is picked out of the bag and not replaced. This is repeated.

Find the probability of...

**(a)** picking one marble of each colour **(b)** not picking a yellow

**(c)** picking two marbles of the same colour.

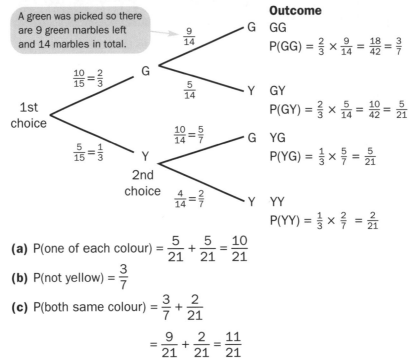

**Outcome**

A green was picked so there are 9 green marbles left and 14 marbles in total.

G GG

$P(GG) = \frac{2}{3} \times \frac{9}{14} = \frac{18}{42} = \frac{3}{7}$

Y GY

$P(GY) = \frac{2}{3} \times \frac{5}{14} = \frac{10}{42} = \frac{5}{21}$

G YG

$P(YG) = \frac{1}{3} \times \frac{5}{7} = \frac{5}{21}$

Y YY

$P(YY) = \frac{1}{3} \times \frac{2}{7} = \frac{2}{21}$

1st choice $\frac{10}{15} = \frac{2}{3}$ G $\frac{9}{14}$ $\frac{5}{14}$

$\frac{5}{15} = \frac{1}{3}$ Y $\frac{10}{14} = \frac{5}{7}$ $\frac{4}{14} = \frac{2}{7}$

2nd choice

**(a)** P(one of each colour) $= \frac{5}{21} + \frac{5}{21} = \frac{10}{21}$

**(b)** P(not yellow) $= \frac{3}{7}$

**(c)** P(both same colour) $= \frac{3}{7} + \frac{2}{21}$

$= \frac{9}{21} + \frac{2}{21} = \frac{11}{21}$

---

**PROGRESS CHECK**

**Without using a calculator**

Use tree diagrams for these questions.

1. This spinner is used in a game. It is spun two times.

   What is the probability of getting...

   **(a)** an even number both times

   **(b)** two numbers that add up to an odd number?

2. The probability of Adam passing his Maths Paper A is 0.6 and Paper B is 0.8

   What is the probability of Adam...

   **(a)** passing both papers?

   **(b)** passing one paper?

   **(c)** failing both papers?

1. (Your tree diagram should have four arms, each with four further arms. There are 16 outcomes in total.) Probability of each outcome = $\frac{2}{4} \times \frac{2}{4} = \frac{1}{16}$

(a) P(even number both times) = $\frac{4}{16} = \frac{1}{4}$

(b) P(two numbers adding up to odd number) = $\frac{1}{16} + \frac{1}{16} + \frac{1}{16} + \frac{1}{16} + \frac{1}{16} + \frac{1}{16} + \frac{1}{16} + \frac{1}{16} = \frac{8}{16} = \frac{1}{2}$

2. (Your tree diagram should have two arms, each with two further arms. There are four outcomes in total.)

P(pass/pass) = $0.6 \times 0.8 = 0.48$; P(fail/fail) = $0.4 \times 0.2 = 0.08$

P(pass/fail) = $0.6 \times 0.2 = 0.12$; P(fail/pass) = $0.4 \times 0.8 = 0.32$

(a) P(passing both) = 0.48 (b) P(passing one) = $0.12 + 0.32 = 0.44$ (c) P(failing both) = 0.08

# Sample GCSE questions

**1** Gary buys a box of coloured pencils in a sale. When he opens the box he finds there are 6 each of red and blue pencils, 4 each of yellow and green pencils and 10 black pencils.

**(a)** Write down the probability of choosing a pencil that is not black. Give as a fraction in its lowest terms. **(2)**

**(b)** Write down the probability of choosing a pencil that is not red or blue. **(1)** Give as a percentage.

**(c)** What is the chance that a pencil chosen at random is:
**(i)** yellow  **(ii)** black or red  **(iii)** orange **(1)**
(Choose from: certain, likely, evens, unlikely, impossible)

**(d)** Mark your answer to part **(c)** with arrows on a probability scale. **(2)**

**There are 20 pencils that are not black** →

(a) Total number in box = 30

$P(\text{not black pencil}) = \frac{20}{30} = \frac{2}{3}$

**Cancel by 6 and multiply by 100** →

(b) $P(\text{not red or blue pencil}) = \frac{18}{30} \times 100 = \frac{3}{5} \times 100 = 60\%$

(c) (i) Unlikely as $\frac{4}{30}$

(ii) Likely as $\frac{16}{30}$

(iii) Impossible, as no orange pencils in box.

(d) (iii) impossible

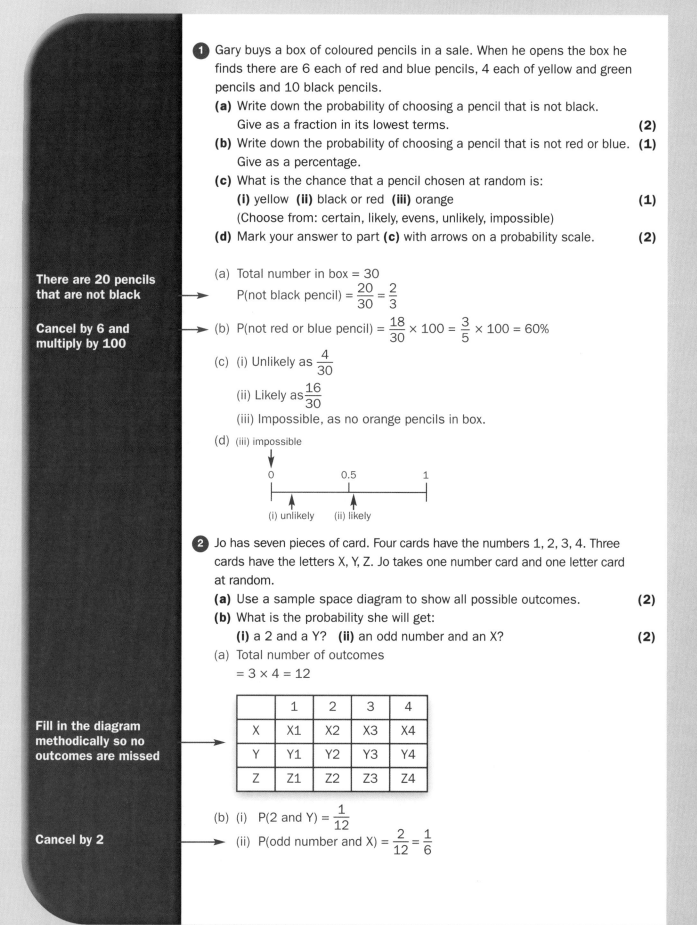

(i) unlikely     (ii) likely

**2** Jo has seven pieces of card. Four cards have the numbers 1, 2, 3, 4. Three cards have the letters X, Y, Z. Jo takes one number card and one letter card at random.

**(a)** Use a sample space diagram to show all possible outcomes. **(2)**

**(b)** What is the probability she will get:
**(i)** a 2 and a Y?   **(ii)** an odd number and an X? **(2)**

(a) Total number of outcomes
= 3 × 4 = 12

**Fill in the diagram methodically so no outcomes are missed** →

|   | 1 | 2 | 3 | 4 |
|---|---|---|---|---|
| X | X1 | X2 | X3 | X4 |
| Y | Y1 | Y2 | Y3 | Y4 |
| Z | Z1 | Z2 | Z3 | Z4 |

(b) (i) $P(2 \text{ and } Y) = \frac{1}{12}$

**Cancel by 2** →

(ii) $P(\text{odd number and } X) = \frac{2}{12} = \frac{1}{6}$

# Exam practice questions

**1** A spinner has numbers 1, 2, 3, 4 on it. It is found to be biased as the probability of getting
a 1 is 0.2, a 3 is 0.3 and a 4 is 0.1
   **(a)** If the spinner was fair, what would the probability of getting a 2 be?
   **(b)** What is the probability of getting a 2 on the biased spinner?
   **(c)** If the spinner is spun 150 times, how many times would the number 3 occur?
   **(d)** Which number is most likely to occur?
   **(e)** What is the probability of getting a 2 or a 3?
   **(f)** What is the probability of getting a 6?                                           **(6)**

**2** Each letter of the word MATHEMATICS is marked on a piece of card. They are put in a bag.
What is the probability of picking out…
   **(a)** a letter A?
   **(b)** a letter C or a letter T?
   **(c)** a vowel or a consonant?                                                          **(4)**

**3** Two boxes of sweets have a combination of milk, plain and white chocolates.
Chococrunch has 10 milk chocolates, 8 plain chocolates and 6 white chocolates.
Chocomix has 8 milk chocolates, 9 plain chocolates and 7 white chocolates.
Rio only likes milk chocolate, Ryan only likes plain chocolate and Wayne only likes white chocolate.
What is the probability of…
   **(a)** Rio picking his favourite from **(i)** Chococrunch? **(ii)** Chocomix?
   **(b)** Ryan not picking his favourite from **(i)** Chococrunch? **(ii)** Chocomix?
   **(c)** Which box would give Wayne a more likely chance of picking his favourite? Why?      **(4)**

**4** The Rich family are going on holiday. The travel agent gives them a choice of three hotels for
their first week, and two hotels for their second week. Because of the facilities offered in
each hotel, the probability of each hotel being chosen by the family in the first week are:
Hotel Parc = 0.2 Hotel Plaza = 0.5 and Hotel Splendide = 0.3
The probability of each hotel being chosen by the family in the second week are:
Hotel Parc = 0.4 Hotel Plaza = 0.6
   **(a)** Complete this tree diagram illustrating all available choices.                     **(5)**

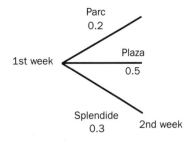

   **(b)** Looking at your results, which combination of hotels is most likely to be chosen by the family?
      If the first combination was not available by the time they actually book, is there another
      combination that they would choose?                                                    **(3)**

# Exam practice answers

## Chapter 1 Number

**1. (a)** Multiples of 5: 15, 25
   **(b)** Factors of 28: 4, 7, 14
   **(c)** Square numbers: 4, 25
   **(d)** $\sqrt[3]{64} = 4$ ($4 \times 4 \times 4 = 64$)
   **(e)** $2^5 = 32$ ($2 \times 2 \times 2 \times 2 \times 2 = 32$)

**2. (a) (i)** $36 = 2^2 \times 3^2$
   **(ii)** $96 = 2^5 \times 3$
   **(b)** LCM of 36 and 96 $= 2^5 \times 3^2 = 288$
   HCF of 36 and 96 $= 2^2 \times 3 = 12$

**3. (a)** $\frac{1}{4} = 25\%$, 0.25 **(b)** $\frac{3}{8} = 37.5\%$, 0.375 **(c)** $\frac{3}{4} = 75\%$, 0.75
   **(d)** $\frac{2}{5} = 40\%$, 0.4

   *Examiner's tip: It is easier to work out the decimal first, by dividing the numerator by the denominator. Multiply the decimal by 100 to find the percentage.*

**4. (a)** February as fraction of a leap year $= \frac{29}{366}$
   (February has 29 days in a leap year; a leap year has 366 days)
   **(b)** 4 weeks as a fraction of a year $= \frac{4}{52} = \frac{1}{13}$
   (Cancel to lowest terms)
   **(i)** $\frac{1}{13} = 0.077$ to 3 d.p. (Divide the numerator by the denominator)
   **(ii)** $0.077 = 7.7 \times 10^{-2}$
   (Standard index form has (number between 1 and 10) $\times 10^n$)

**5. (a)** Sale price $= £(107.95 - 20) = £87.95$
   Percentage decrease $= \frac{20}{107.95} \times 100 = 18.5\%$
   **(b)** Price £87.95 including VAT at 15%
   Price without VAT $= \frac{87.95}{1.15} = £76.48$
   Price with 17.5% VAT $= £76.48 \times 1.175$
   $= £89.86$

   *Examiner's tip: Always ask yourself if the answer makes sense. Cost will increase if VAT is increased.*

**6.** There are two ways to compare the accounts:
   **(i)** interest earned and **(ii)** final amount.
   **(i)** Interest earned
   Simple interest account:
   interest over 2 years $= £12\,500 \times 2 \times 0.0375$
   $= £937.50$

   Compound interest account:
   1st year interest $= 12\,500 \times \frac{4.25}{100} = £531.25$
   2nd year interest $= (12\,500 + 531.25) \times \frac{4.25}{100}$
   $= £553.83$

   total interest over 2 years $= £531.25 + £553.83$
   $= £1085.08$

   ∴ Adam should use the 2 year compound interest account as it pays £147.58 more interest.

**(ii)** Final amount
   Simple interest account:
   final amount = original amount + total interest
   $= £12\,500 + £937.50$ (working in **(i)**)
   $= £13\,437.50$

   Compound interest account:
   final amount $= £12\,500(1 + 0.0425)^2$ (see page 29 for formula)
   $= £12\,500 \times 1.0425^2 = £13\,585.08$

**7. (a)** Coldest place is Reykjavik.
   **(b) (i)** London to Prague: -2°C
   **(ii)** Reykjavik to Zurich: +4°C
   *Examiner's tip: You are asked for difference so include + to show increase.*
   **(iii)** Stockholm to Milan: +13°C
   *Examiner's tip: Always take care when working with negative numbers.*

**8. (a)** £10 buys €12 ⇒ £50 buys €($5 \times 12$) = €60
   *Examiner's tip: You can work out the rate for £1 instead*
   **(b)** €25 buys £15 ⇒ €5 buys £($15 \div 5$) = £3 ⇒ €60 buys £($3 \times 12$) = £36
   **(c)** Iqra lost £14 by having to exchange back from euros to pounds.
   **(d)** Percentage loss $= \frac{14}{50} \times 100 = 28\%$

## Chapter 2 Algebra

**1. (a)** $4a - b$ **(b)** $2y - x$
   **(c)** $p - 2p^2 - q^2 + pq$ or $p(1 - 2p) - q(q - p)$
   *Examiner's tip: Collect like terms after expanding brackets. Take care with signs.*

**2. (a) (i)** $v^2 = u^2 + 2as \Rightarrow v^2 - u^2 = 2as \Rightarrow a = \frac{v^2 - u^2}{2s}$
   **(ii)** $x = \sqrt{\frac{a}{a+b}}$ (Square both sides of formula to eliminate square root)
   $x^2 = \frac{a}{a+b} \Rightarrow x^2(a + b) = a \Rightarrow ax^2 + bx^2 = a$
   $\Rightarrow ax^2 - a = -bx^2 \Rightarrow a(x^2 - 1) = -bx^2 \Rightarrow a = \frac{-bx^2}{x^2 - 1}$
   or $\frac{bx^2}{1 - x^2}$
   (Collect all terms including $a$ on the same side)
   **(b) (i)** $5 > x - 3 \Rightarrow 5 + 3 > x \Rightarrow 8 > x$
   **(ii)** $2(3y + 4) < 3 \Rightarrow 6y + 8 < 3 \Rightarrow 6y < 3 - 8$
   $\Rightarrow 6y < -5 \Rightarrow y < \frac{-5}{6}$
   (Expand brackets first)
   **(iii)** $4(x - 3) \leqslant 18 + 2x \Rightarrow 4x - 12 \leqslant 18 + 2x$
   $\Rightarrow 4x - 2x \leqslant 18 + 12 \Rightarrow 2x \leqslant 30 \Rightarrow x \leqslant 15$

**3. (a)** 0, 2, 6, 12, 20.....
   (Differences between terms are 2, 4, 6. The next difference must be 8, so next term = 12 + 8 = 20)
   **(b)** $n$th term $= n^2 - n$ or $n(n - 1)$

| Term | 1 | 2 | 3 | 4 | 5 | $n$th term |
|---|---|---|---|---|---|---|
| Calculation | $1^2 - 1$ | $2^2 - 2$ | $3^2 - 3$ | $4^2 - 4$ | $5^2 - 5$ | $n^2 - n$ |
| Result | 0 | 2 | 6 | 12 | 20 | |

**4. (a)** $q(p - m) = \frac{1}{2}(3 - (-1)) = \frac{1}{2}(3 + 1) = \frac{1}{2} \times 4 = 2$

*Examiner's tip: Show your substitution. You will score method marks, even if you make a mistake calculating the final answer.*

**(b)** $(m + n + p)^2 = (-1 + 2 + 3)^2 = 4^2 = 16$

**(c)** $\frac{mn + nq + p}{n} = \frac{-2 + 1 + 3}{2} = \frac{2}{2} = 1$

**5.** (*Use p = pen and c = pencil to form two simultaneous equations. Change money to pence, it is easier for calculations*)

Jonah buys 3 pens and 5 pencils for £3.25

    equation 1    $3p + 5c = 325$

His sister buys 2 pens and 3 pencils for £2.05

    equation 2    $2p + 3c = 205$

(*Eliminate p by multiplying equations to make coefficients the same, then subtract*)

    equation 3    $6p + 10c = 650$

    equation 4    $6p + 9c = 615$

equation 4 – equation 3:

$6p - 6p + 10c - 9c = 650 - 615$

                     $c = 35$

Substitute $c = 35$ in equation 1:

$3p + 5(35) = 325$

 $3p + 175 = 325$

       $3p = 325 - 175$

       $3p = 150$

         $p = 50$

A pencil costs 35p and a pen costs 50p.

*Examiner's tip: Always give answers in terms of the question.*

**6. (a)**  $3(3 + 2y) = y - 16$   (Expand brackets first)

     $9 + 6y = y - 16$   (Collect terms)

       $6y - y = -16 - 9$

          $5y = -25$

           $y = -5$

**(b)** $x^2 + 4x - 12 = 0$

   $(x + 6)(x - 2) = 0$    (Find factors of 12)

   Either $(x + 6) = 0 \Rightarrow x = -6$

   or $(x - 2) = 0 \Rightarrow x = 2$ (Remember there are usually two solutions to a quadratic equation)

**(c)** $16a^2 - 36 = 0$     (Difference of two squares)

   $(4a - 6)(4a + 6) = 0$

   Either $(4a - 6) = 0 \Rightarrow a = \frac{6}{4} = \frac{3}{2} = 1\frac{1}{2} = 1.5$

   or $(4a + 6) = 0 \Rightarrow a = -\frac{6}{4} = -\frac{3}{2} = -1\frac{1}{2} = -1.5$

**(d)** $\frac{x + 12}{7} = \frac{x - 4}{3} \Rightarrow 3(x + 12) = 7(x - 4)$

   $\Rightarrow 3x + 36 = 7x - 28 \Rightarrow 4x = 64 \Rightarrow x = 16$

**7. (a)** D     (Minus sign means graph is inversed)

**(b)** C     ($x$ is denominator, so reciprocal function)

**(c)** B     (-5$x$ moves curve along +$x$-axis)

**(d)** A     (+3 moves curve up +$y$-axis)

## Chapter 3 Geometry

**1.** Use Pythagoras' theorem

   length of ladder $= \sqrt{2.75^2 + 1^2}$

                   $= 2.93$m

Use cosine ratio:

angle between ladder and floor $= \cos^{-1} \frac{1}{2.93} = 70°$

**2.** ∠ DEX = 60° (Angle sum of Δ = 180°)

   ∴ Δ DEX is equilateral (3 equal angles)

   ∴ XD = EX

   The diagonals of a rectangle bisect each other.

   ∴ FX = EX

   ∴ Δ EFX is an isosceles triangle.

**3. (a)** Exterior angle = 360° ÷ 6 = 60° (Exterior angle of polygon = 360 ÷ number of sides)

   Interior angle = 180° – 60° = 120° (Angles on a straight line add up to 180°)

**(b)** Angles at one point add up to 360°

   360° ÷ 120° = 3 ∴ 3 hexagons fit together at a point.

**4. (a)** ∠ $x$ = 65° (Alternate angles); ∠ $y$ = 30° (Alternate angles);

   ∠ $z$ = 65° + 30° = 95° (Exterior angle of Δ = sum of 2 opposite interior angles)

**(b)** Use sine ratio to find perpendicular height of Δ RST.

   perpendicular height = 8.5 × sin 30° = 4.25cm

**5.** Use sine rule to find missing angle:

   $\frac{\sin A}{15} = \frac{\sin 53°}{19} \Rightarrow \sin A = \frac{15 \sin 53°}{19}$

   ∠ A = 39° (use $\sin^{-1}$) so ∠ B = 88°

   Use cosine rule to find missing side:

   $b^2 = 15^2 + 19^2 - 2(15 \times 19) \cos 88°$

      $= 225 + 361 - 570 \cos 88°$

      $= 566$

   missing side $= \sqrt{566} = 23.8$cm to 3 s.f.

   (Do not forget to square root)

**6.** ∠ BTR = 48° and ∠ BVR = 35° (Alternate angles as ground parallel to line of sight from top of building)

   Use tan ratio to find BT and BV.

   BT $= \frac{87}{\tan 48°} = 78.3$m; BV $= \frac{87}{\tan 35°} = 124.2$m

   ∴ VT = BV – BT = 124.2 – 78.3 = 45.9m

          = 46m (to nearest m)

**7.** The pairs of congruent triangles are: Δ CYF and Δ DYE; Δ DEF and Δ CFE; Δ FCD and Δ EDC

   (*Look for equal sides and angles*)

**8.** ∠ DBA = 57° (Angles in same segment of circle)

   ∴ ∠ CAB = 82° – 57° = 25° (Exterior angle of Δ = sum of 2 opposite interior angles)

**Chapter 4 Measures**

1. **(a)** Area = (5.8 × 6.7) + 2(5.8 × 2.5) + 2(6.7 × 2.5)
      = 38.86 + 29 + 33.5
      = 101.36m²
   **(b)** Number of cans = 101.36 ÷ 15 = 6.76
      ∴ 7 cans must be bought
   **(c)** Cost of cans = 7 × £11.60 = £81.20

2. **(a)** Radius of beaker = $\frac{1}{2}$ diameter = 5.5 ÷ 2 = 2.75cm
      capacity of beaker = $\pi r^2 h = \pi \times 2.75^2 \times 10$
         = 237.6cm³ to 1 d.p.
      jug holds 2.5l = 2.5 × 1000 = 2500cm³
         (1 litre = 1000ml or 1000cm³)
      number of times beaker can be filled by jug
         = 2500 ÷ 237.6 = 10.5
   **(b)** Area of label = circumference of base × height
      (curved surface of can)
      circumference of base = $\pi d = \pi \times 5.5$
      (There is no need to work this out. Leave in terms of $\pi$)
      area of label = $\pi \times 5.5 \times 10$ = 172.8cm² to 1 d.p.
   **(c)** Total surface area of the can = area of curved
      surface + 2 circular ends
      area of 2 ends = $2 \times \pi \times 2.75^2$ = 47.5cm²
      total surface area of the can = 172.8 + 47.5
         = 220cm² to 3 s.f.

3. PQH is a right-angled
   triangle with PQ as the
   hypotenuse.

   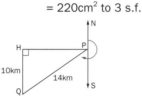

   $PQ^2 = QH^2 + PH^2$ using Pythagoras' theorem giving
   $PH^2 = PQ^2 - QH^2 \Rightarrow PH = \sqrt{(14^2 - 10^2)} = \sqrt{96}$
         PH = 9.8km
   bearing of Q from P = 180° + ∠ QPS
   ∠ QPS = ∠ PQH (Alternate angles as QH and PS are
   parallel)
   using cosine ratio in Δ PQH: ∠ QPS = $\cos^{-1} \frac{10}{14}$ = 44° to
   nearest degree   (Use shift cos to get cos⁻¹)
   bearing of Q from P = 180° + 44° = 224°
   *Examiner's tip: Remember, for right-angled triangles, use
   Pythagoras' theorem to calculate sides; use trigonometric
   ratios to calculate angles and sides.*

4. **(a)** $\frac{1}{2}$ or 0.5
   **(b)** $\frac{3}{4}$ or 0.75
   **(c)** $1\frac{1}{4}$ or $\frac{5}{4}$ or 1.25
   **(d)** $1\frac{1}{2}$ or $\frac{3}{2}$ or 1.5
   **(e)** 2

5. Volume of one sphere = $\frac{4}{3}\pi r^3 = \frac{4}{3} \times \pi \times 4^3$ = 268.1cm³
   volume of three spheres = 3 × 268.1 = 804cm³
   height of tube = 3 × diameter of sphere = 3 × 8 = 24cm
   volume of tube = $\pi r^2 h = \pi \times 4^2 \times 24$ = 1206cm³
   ∴ volume of tube not filled by spheres = 1206 – 804
   = 402cm³

6. **(a)** Perimeter = OA + OB + arc AB   (Use Pythagoras to
                                        find OA and OB)
      OA = OB = $\sqrt{(15^2 + 10^2)} = \sqrt{325}$ = 18cm
      quadrant of circle so arc AB = $\frac{1}{4}(2\pi r) = \frac{1}{4} \times 2 \times \pi \times 18$
      = 28.3cm
      perimeter = (2 × 18) + 28.3 = 64.3cm
      Area of the whole shape = $\frac{1}{4}(\pi r^2) = \frac{1}{4} \times \pi \times 18^2$
         = 254cm² to 3 s.f.
   **(b)** Area of the boat = area of quadrant – area of triangle
      = 254 – ($\frac{1}{2} \times 20 \times 15$) = 254 – 150 = 104cm²

**Chapter 5 Statistics**

1. **(a)** **Any three suitable questions, e.g.**
      What is your age and gender?
      M ☐  F ☐  10–12 ☐  13–15 ☐  16–18 ☐
      How much would you pay to download a single piece
      of music? (Plus response boxes)
      What type of music do you prefer? (Plus response boxes)
      *Examiner's tip: Giving response boxes usually gains a mark.*
   **(b)** No, it is not a suitable sample. It is biased to one
      narrow age group.

2. Total girls' marks = 1130; total number of girls = 26
   ∴ Mean of girls' marks = $\frac{1130}{26}$ = 43.5
   Median of girls' marks is halfway between 13th and
   14th result = 44
   Total boys' marks = 985; total number of boys = 22
   ∴ Mean of boys' marks = $\frac{985}{22}$ = 44.8
   Median of boys' marks is halfway between 11th and
   12th result = 44
   ∴ Although medians are same, the average mark for
   the boys is higher than that of the girls.
   *Examiner's tip: You need to give evidence to gain full
   marks. It is usually best to compare averages or range.*

3. **(a)** *(You must work out the product of the midpoint and
      the frequency for each group to include all results)*

| Speed (s mph) | Frequency (f) | Midpoint (m) | (fm) |
|---|---|---|---|
| 40 ⩽ s < 50 | 4 | 45 | 4 × 45 = 180 |
| 50 ⩽ s < 60 | 25 | 55 | 25 × 55 = 1375 |
| 60 ⩽ s < 70 | 33 | 65 | 33 × 65 = 2145 |
| 70 ⩽ s < 80 | 35 | 75 | 35 × 75 = 2625 |
| 80 ⩽ s < 90 | 3 | 85 | 3 × 85 = 255 |
| Total | Σf = 100 | | Σfm = 6580 |

Estimate of the mean speed of cars = $\frac{\Sigma fm}{\Sigma f} = \frac{6580}{100}$
         = 65.8mph
The average driver is driving within the speed limit of
70mph.

**(b)**

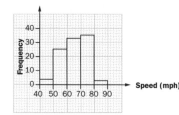

**4.** 0–25: slight increase in B, 26–50: declines in B (people in this age range may move for their work); 51–75: increase in B (people may retire to the area) and 75+: greater in B (people living longer).

**5. (a)**

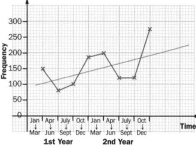

**(b)** See graph

**(c)** Estimated number of subscriptions bought in the next three months = 200+ (Use the trend line)

## Chapter 6 Probability

**1. (a)** 0.25 or $\frac{1}{4}$ (Equal chance for each number)

**(b)** Probability of getting a 2 on the biased spinner = 1 − (0.2 + 0.3 + 0.1) = 0.4 (Total probability = 1)

**(c)** If the spinner is spun 150 times, the number 3 would occur 45 times. (Using relative frequency)

**(d)** 2 is most likely to occur. (Highest probability)

**(e)** P(getting a 2 or a 3) = 0.4 + 0.3 = 0.7 (Use OR rule)

**(f)** P(getting a 6) = 0 (Impossible)

**2. (a)** P(picking out A) = $\frac{2}{11}$

**(b)** P(picking out C or T) = $\frac{1}{11} + \frac{2}{11} = \frac{3}{11}$

**(c)** P(picking out a vowel or a consonant) = 1 (Certain as every letter is a vowel or a consonant)

**3. (a) (i)** $\frac{5}{12}$ **(ii)** $\frac{1}{3}$

**(b) (i)** $\frac{2}{3}$ **(ii)** $\frac{5}{8}$

**(c)** Chocomix because there is a greater proportion of white chocolates (Probability = $\frac{7}{24}$) than in Chococrunch (Probability = $\frac{6}{24}$).

**4. (a)** *Examiner's tip: Remember to multiply along branch (AND rule) and add when looking at more than one outcome (OR rule). Put all working down. You will lose marks if you do not. Draw the tree in pencil. It will be easier to correct a mistake. When you are sure about your tree, you can use ink to write in the probabilities.*

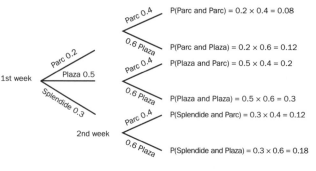

**(b)** The highest probability is P(Plaza and Plaza) = 0.3 ∴ combination of hotels most likely to be chosen by the family is two weeks in Hotel Plaza.
The next highest probability is P(Plaza and Parc) = 0.2 ∴ if the first combination is not available by the time of booking, the other combination likely to be chosen by the family is one week in Hotel Plaza and one week in Hotel Parc.

# Further exam practice

The following papers have been written to provide you with an opportunity to practise a formal type of examination paper before taking the actual GCSE exams in school. The format and layout of the questions are very similar to those that will be in the exams although the length of each paper and the marks available will differ depending on which exam board specification you are studying. You should consult the exam board specification or your teacher so that you know what to expect in the exam. These papers will help you to prepare and develop skills to enable you to approach the actual test confidently and achieve the best that you can. You should allow 1hr 45 mins to complete each paper.

---

**Instructions**

- Use black ink or black ball-point pen. Pencils may be used for graphs and diagrams only.
- Answer all questions in the spaces provided on the question paper.
- Show all your working, as method marks may be given.
- Put a line through any rough working you do not wish to be marked.

---

**Information**

- The number of marks is given in brackets [ ] at the end of each question or part question. Use this as a guide to the length of answer required.
- The total marks for each paper is 100.

---

**Advice**

- Read each question carefully before you start your answer.
- Attempt all questions.
- Show all stages of your working and state units.
- Keep track of the time.
- Check your answers, if possible, when you have finished.

# Useful Formulae

**Area of trapezium** = $\frac{1}{2}(a + b)h$

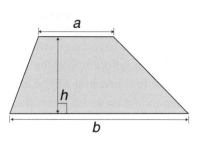

**Volume of prism** = area of cross section × length

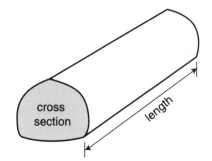

**Volume of sphere** = $\frac{4}{3}\pi r^3$

**Surface area of sphere** = $4\pi r^2$

**Volume of cone** = $\frac{1}{3}\pi r^2 h$

**Curved surface area of cone** = $\pi r l$

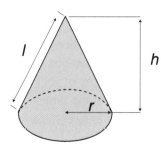

**In any triangle ABC**

**Area of triangle** = $\frac{1}{2}ab \sin C$

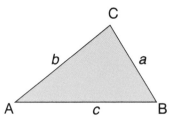

**Sine rule** $\dfrac{a}{\sin A} = \dfrac{b}{\sin B} = \dfrac{c}{\sin C}$

**Cosine rule** $a^2 = b^2 + c^2 - 2bc \cos A$

**The quadratic equation**
The solutions of $ax^2 + bx + c = 0$, where $a \neq 0$, are given by $x = \dfrac{-b \pm \sqrt{(b^2 - 4ac)}}{2a}$

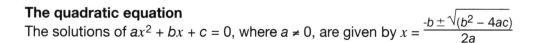

## Higher Tier

## Paper 1

**Calculators must not be used for this paper**

**1. (a) (i)** Give 70 and 84 as the product of their prime factors.

[2 marks]

**(ii)** Find the lowest common multiple and the highest common factor of 70 and 84.

[2 marks]

**(b)** Find $\frac{1}{5}$ of $5^6$. Give the answer as a power.

[1 mark]

**(c)** Work out:

**(i)** $1\frac{3}{5} \div \frac{1}{6}$

[1 mark]

**(ii)** $2\frac{3}{7} - 1\frac{2}{3}$

[1 mark]

**(d)** Write $0.6\dot{1}$ as a fraction.

[2 marks]

**2. (a)** Estimate the value of $\dfrac{43 \times 261}{197}$

[2 marks]

**(b)** Simplify $(p^6)^{\frac{1}{2}} \times 5pq$

[2 marks]

**(c)** Factorise $3mn^2 - 9mn + 6m^2$

[2 marks]

**3.** Correct to 3 significant figures and then express in standard index form:

**(a)** 0.000 006 130 2

[2 marks]

**(b)** 52 480 000

[2 marks]

**4. (a)** Solve the inequality $3x - 4 > 11 - 2x$

_____

_____ [2 marks]

**(b)** Show your solution on this number line.

-1  0  1  2  3  4  5  6  7 [1 mark]

**5. (a)** Find the value of $3x^2 + 2y^3$ when $x = 2$ and $y = -1$

_____ [2 marks]

**(b)** Which of these fractions is closest to $\frac{1}{2}$? Show all your working.

$\frac{1}{4}$   $\frac{2}{5}$   $\frac{3}{10}$   $\frac{9}{20}$   $\frac{3}{50}$

_____

_____

_____ [2 marks]

**6.** Naomi and Gary have just moved house. They decide to tile the bathroom.
They choose patterned wall tiles, which are made in two colours and many different sizes.
Each tile is made of small squares. Some of the squares are a light shade and some are a
dark shade. Here are the smallest three sizes of tiles:

**Tile 1**        **Tile 2**        **Tile 3**

**(a)** Naomi and Gary need to know how to work out the number of light and dark squares in a
larger tile. Find an expression for **(i)** the number of light squares and **(ii)** the number of dark
squares in the $n$th tile.

_____

_____

_____

**(i)** _____  **(ii)** _____ [4 marks]

**(b)** How many light and dark squares would tile 8 have?

_____

_____ [2 marks]

**(c)** Each small square on the tile has a side measuring 30mm.
Naomi and Gary decide they want to use a tile with a side measuring 21cm. Which tile is this?

_____

_____ [2 marks]

**(d)** The cost of the tiles increases in the same ratio as the measurements of their sides.
If tile 1 costs £1.50, how much will Naomi and Gary pay for their chosen tile?

........................................................................................................................................................

........................................................................................................................................................

Cost = £ ........................................................ [3 marks]

**7.** **(a)** Make $p$ the subject of this formula:

$$\frac{2m(6 + p)}{5 - p} = 3$$

........................................................................................................................................................

........................................................................................................................................................

........................................................................................................................................................

........................................................................................................................................................ [4 marks]

**(b)** Solve this equation: $y^2 + 2y - 15 = 0$

........................................................................................................................................................

........................................................................................................................................................

........................................................................................................................................................ [4 marks]

**8.** Mike needs to cut some wood and wants to hire an electric saw.
This graph shows the hire cost ($H$) for a number of days ($d$).

**(a)** Write down the formula for the cost of hire.

........................................................................................................................................................ [1 mark]

**(b)** Why does the line not go through zero?

........................................................................................................................................................ [1 mark]

**(c)** How much will it cost to hire the saw for 5 days?

........................................................................................................................................................

........................................................................................................................................................ [2 marks]

**(d)** How many days hire will Mike get for £40?

........................................................................................................................................................

........................................................................................................................................................ [2 marks]

**9.** A circle has centre O and radius of 5cm. A chord AB subtends an angle of 100° at the centre.

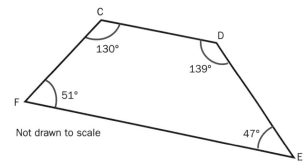

Not drawn to scale

**(a)** Calculate the length of AB.

........................................................................................

........................................................................................

[2 marks]

**(b)** Another chord, BC, is drawn from B to meet the end of the diameter AC. Calculate ∠ACB.

........................................................................................

........................................................................................

[2 marks]

**10.** A park has four paths CD, DE, EF and FC in the shape of a quadrilateral.

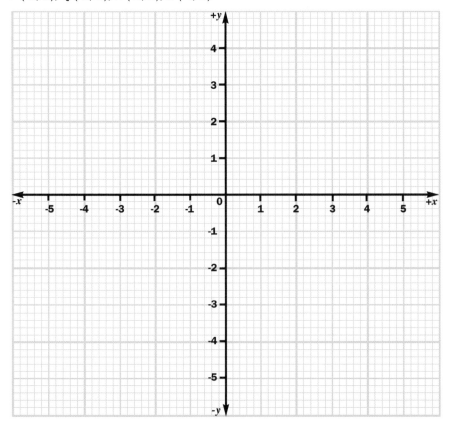

Not drawn to scale

The park ranger would like to make a circular children's playground in the area of the paths. She would need to enclose it by installing a fence. She thinks the fence could be a circle with C, D, E and F all on the circumference. Is she right? Give reasons for your answer.

........................................................................................

........................................................................................

[2 marks]

**11. (a)** Plot these coordinates on the graph paper and join them to form a quadrilateral.
P (-3, 3), Q (-2, 3), R (-2, 1), S (-4, 2)

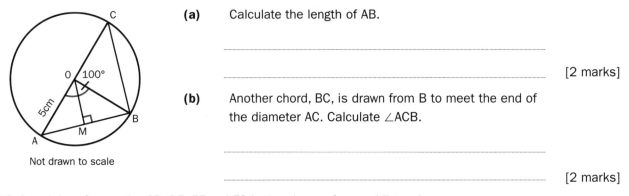

[2 marks]

**(b)** Rotate PQRS anticlockwise 90° about O. Label this image A. [2 marks]

**(c)** Reflect image A in the line $y = -x$ to form image B.
What are the coordinates of the vertices of image B?

........................................................................................................................

........................................................................................................................ [4 marks]

**(d)** Would the column vector $\begin{pmatrix} -4 \\ 0 \end{pmatrix}$ move image B back to the original position?
Give a reason for your answer.

........................................................................................................................ [1 mark]

**12.** Find the coordinates of the points of intersection of a line $3x + 2y = 6$ and a curve $y = x^2 - 5x + 6$
graphically.

........................................................................................................................

........................................................................................................................

........................................................................................................................

........................................................................................................................

........................................................................................................................ [6 marks]

Points of intersection: ........................... and ........................... [2 marks]

**13.** K, L and M are three villages. KL = 18km, LM = 10km. The bearing of L from K is 080° and of M
from L is 135°. Jacob has to travel from L to K as part of an orientation exercise.
**(a)** Work out the bearing he will have to take.

........................................................................................................................

........................................................................................................................ [2 marks]

**(b)** A group from his school are walking a longer trail from K to M via L.
Construct a scale drawing of their journey. What is the actual direct distance and bearing of M from K?
Use a scale of 1cm = 3km.

...........................................................................................................................................................................

...........................................................................................................................................................................

........................................................................................................................................................... [6 marks]

**14.** ABCD is a rhombus. M is the point of intersection of the diagonals and N is the midpoint of $\overrightarrow{AB}$.
**a**, **b** and **c** are vectors.

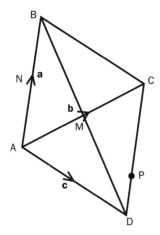

**(a)** Find $\overrightarrow{CD}$, $\overrightarrow{AM}$ and $\overrightarrow{MN}$.

...........................................................................................................................................................................

...........................................................................................................................................................................

...........................................................................................................................................................................

........................................................................................................................................................... [3 marks]

**(b)** P is a point on $\overrightarrow{CD}$, so that $\overrightarrow{CP} : \overrightarrow{PD} = 2 : 1$. Prove that $\overrightarrow{MP} = \dfrac{4\mathbf{c} - \mathbf{b}}{6}$

...........................................................................................................................................................................

...........................................................................................................................................................................

........................................................................................................................................................... [3 marks]

15. (a) Two walls, each 5.5m long, are built to meet at an angle as shown below.

   (i) Measure the angle between them.

   (ii) Measure a wall and work out the scale that has been used in this drawing.

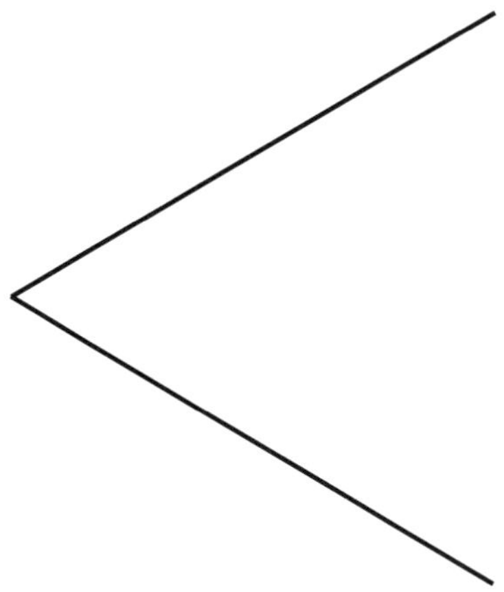

   .................................................................................................................................... [1 mark]

   (b) A gardener plants a line of seedlings equidistant from the walls, so that they are sheltered from the wind. Construct the locus of these seedlings on the diagram above, showing all construction lines. What line have you drawn?

   .................................................................................................................................... [2 marks]

   (c) Measure the angle between the line of seedlings and a wall.

   .................................................................................................................................... [1 mark]

   (d) Esther wants her dog to be able to run in this part of the garden, but her dad is worried the dog will damage the plants. She promises to restrict the dog by tethering him by a chain to a hook on one wall. The length of the chain is 2m. Mark a point on each wall to show where a hook could be placed, so the dog cannot harm the seedlings. Shade in the area the dog may use. [2 marks]

**16. (a)** This is a sketch of $y = f(x)$

Sketch the curve $y = f(x) - 3$ using the same axes.

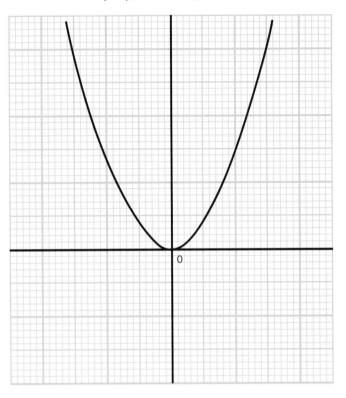

What are the coordinates of the point where $y = f(x) - 3$ cuts the $y$-axis?

$x =$ ........................   $y =$ ........................                                    [3 marks]

**(b)** This is a sketch of $y = g(x)$

Sketch the curve $y = -g(x)$ using the same axes.

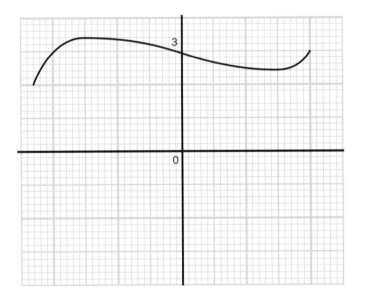

What are the coordinates of the point where $y = -g(x)$ cuts the $y$-axis?

$x =$ ........................   $y =$ ........................                                    [3 marks]

## Higher Tier

## Paper 2

**Calculators may be used for this paper**

(N.B. If your calculator does not have a π button, take the value of π to be 3.142 unless told otherwise.)

1.  Solve these equations:
    **(a)** $5y + 6 = 18$

    ...........................................................................................................................................................

    ........................................................................................................................................................... [2 marks]

    **(b)** $2(x - 1) = 3(5 - 2x)$

    ...........................................................................................................................................................

    ...........................................................................................................................................................

    ...........................................................................................................................................................

    ........................................................................................................................................................... [3 marks]

    **(c)** $z + 3 = \dfrac{4(2 - z)}{3}$

    ...........................................................................................................................................................

    ...........................................................................................................................................................

    ...........................................................................................................................................................

    ........................................................................................................................................................... [3 marks]

2.  A carton, of volume 600cm³, has length and width each 6cm more than its height. Use trial and improvement to find the height ($h$) correct to 1 decimal place. Show all trials and outcomes.

    $h =$................................... [3 marks]

3.  **(a)** Phil and Fiona are buying a new car. It costs £12 599 inclusive of VAT at 17.5%.
        What is the actual value of the car (without VAT)?

    ...........................................................................................................................................................

    ........................................................................................................................................................... [2 marks]

**(b)** Before 1st January, VAT was charged at 15%. What would the total cost of the car have been?

........................................................................................................................

........................................................................................................................  [2 marks]

**(c)** What is the percentage increase in total price after 1st January?

........................................................................................................................

........................................................................................................................  [2 marks]

**4.** BD is parallel to AE
∠CEA is a right angle
∠ADE = 56°; ∠ACE = 28°

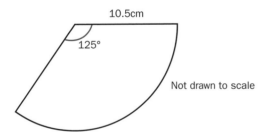

Not drawn to scale

**(a)** Find AE and CD.

........................................................................................................................

........................................................................................................................

........................................................................................................................

........................................................................................................................  [3 marks]

**(b)** Show that Δ ACD is isosceles.

........................................................................................................................

........................................................................................................................  [2 marks]

**(c)** What kind of triangles are Δ BCD and Δ ACE? Use your answer to find BD.

........................................................................................................................

........................................................................................................................

........................................................................................................................  [2 marks]

**5.** This sector of a circle is the net of a cone used for holding chips.
The circle has radius 10.5cm and an angle of 125° at its centre

10.5cm

125°

Not drawn to scale

Work out the vertical height of the cone correct to 3 s.f.

........................................................................................................................

........................................................................................................................

........................................................................................................................  [3 marks]

**6.** Miriam and Dave buy their children a sandpit for the garden.
It is in the shape of a cuboid measuring 120cm × 120cm × 28cm.

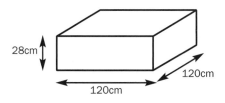

Sandpit sand is sold by two different stores.

| Store | Price per bag | Delivery |
|---|---|---|
| **BuyRite** | £5.99 (25kg) | free |
| **CostOK** | £2.99 (15kg) | £5 |

Miriam and Dave need enough sand to fill the sandpit up to 20cm depth.

**(a)** Find the volume of sand needed in m³.

............................................................................................................................................

............................................................................................................................................ [2 marks]

**(b)** 1m³ of sand weighs 1.6 tonnes. Work out the weight of sand needed in kg.

............................................................................................................................................

............................................................................................................................................ [2 marks]

**(c)** How many bags of sand should they buy? Which store would give better value?

............................................................................................................................................

............................................................................................................................................

............................................................................................................................................

............................................................................................................................................ [3 marks]

**7.** A cone has base radius 10cm and perpendicular height 12cm.
A small cone with half the radius and half the height is sliced off the top.

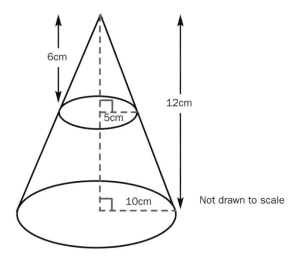

Not drawn to scale

**(a)** Find the volume of the remaining frustum. Give the answer in terms of π.

_____

_____

_____

_____ [3 marks]

**(b)** The frustum has the same volume as a cone with a perpendicular height of 30cm.
Calculate the radius and total surface area of this cone.

_____

_____

_____

_____ [3 marks]

**8.** This diagram is of the grass area inside
a running track, which has two identical
semi-circular ends.

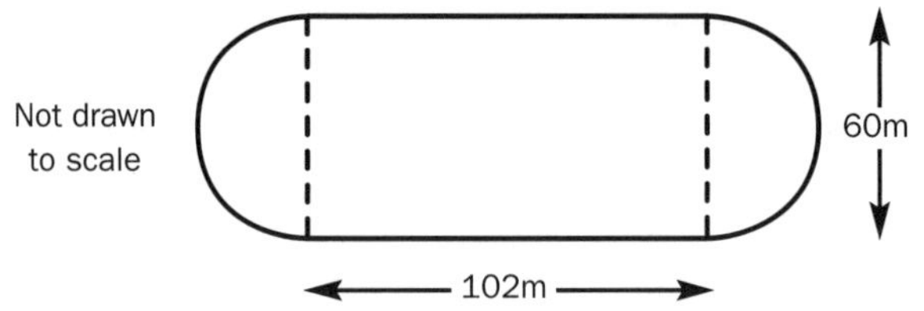

Not drawn
to scale

102m

60m

**(a)** Kirsty runs next to the grass. How far does she run on one lap of the track?

_____

_____

_____ [2 marks]

**(b)** Each individual lane is 1.25m wide. If there are six lanes, how wide is the total running track?

_____ [1 mark]

**(c)** Sophie runs a lap on the far edge of lane six. How much further does she run than Kirsty?

_____

_____

_____ [2 marks]

**(d)** Calculate the area of turf needed for the grass area. Give the answer to the nearest 1000m$^2$.

_____

_____

Area = _____ m² [3 marks]

**9.** This table shows information about the ages of the people living in a block of flats.

| Age (x years) | $20 \leqslant x < 30$ | $30 \leqslant x < 40$ | $40 \leqslant x < 50$ | $50 \leqslant x < 60$ | $60 \leqslant x < 70$ |
|---|---|---|---|---|---|
| Frequency (f) | 18 | 25 | 28 | 11 | 8 |

**(a)** Which class interval contains the median and which is the modal class?

_____ [2 marks]

_____

**(b)** Draw a frequency polygon to illustrate the information.

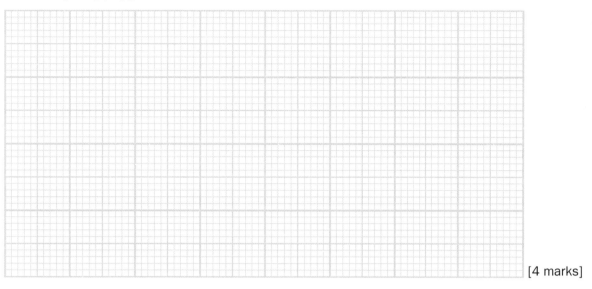

[4 marks]

**(c)** In order to decide on improvements to the flats, a stratified sample of 30 residents is to be taken. Work out the number of people from each group in the sample.

........................................................................................................................................................................................

........................................................................................................................................................ [2 marks]

**10.** A factory tests how long batteries last in a toy car.

| Time ($t$ hours) | Frequency ($f$) | Cumulative frequency |
|---|---|---|
| $8 \leqslant t < 10$ | 5 | |
| $10 \leqslant t < 12$ | 10 | |
| $12 \leqslant t < 14$ | 18 | |
| $14 \leqslant t < 16$ | 13 | |
| $16 \leqslant t < 18$ | 4 | |

**(a)** Complete the cumulative frequency table. [2 marks]

**(b)** Draw a cumulative frequency graph illustrating this data.

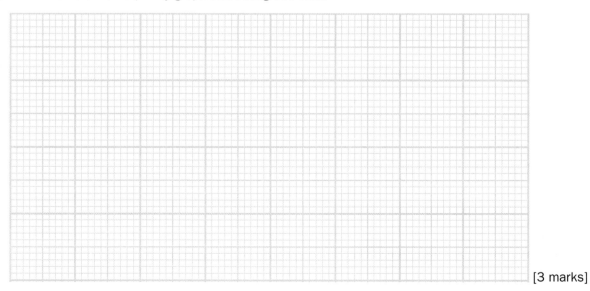

[3 marks]

(c) Use the graph to estimate the median life of a battery and the inter-quartile range.

..................................................................................................................................................................

Median = ...................................................          Inter-quartile range = ................................................          [2 marks]

(d) What percentage of batteries last more than 15 hours?

..................................................................................................................................................................          [2 marks]

**11.** A bag contains 10 small balls. 6 of the balls are red and 4 of the balls are blue.
A ball is chosen at random and then replaced.
A second ball is then chosen at random.

(a) Draw a tree diagram showing all possible outcomes, with probabilities marked.

[3 marks]

(b) Find the probability of choosing 2 blue balls.

..................................................................................................................................................................          [1 mark]

**12.** A child's building brick is in the shape of a prism, length 9cm.
The cross-section of the prism is a 60° sector of a circle, radius 3cm.

(a) Calculate the volume of the brick.

..................................................................................................................................................................

..................................................................................................................................................................          [3 marks]

**(b)** How many bricks can be fitted together to form a cylinder?

............................................................................................................................................................ [1 mark]

**(c)** The curved side of each brick is painted red. What is the area of red on the cylinder?

............................................................................................................................................................

............................................................................................................................................................

............................................................................................................................................................ [2 marks]

**13.** Calculate ∠ABC, ∠BEC and ∠ABE in this diagram.

............................................................................................................................................................

............................................................................................................................................................

............................................................................................................................................................

............................................................................................................................................................

............................................................................................................................................................

............................................................................................................................................................ [3 marks]

**14.** Two dice are thrown at the same time. Use a sample space diagram to show all possible outcomes.

|   | 1 | 2 | 3 | 4 | 5 | 6 |
|---|---|---|---|---|---|---|
| 1 |   |   |   |   |   |   |
| 2 |   |   |   |   |   |   |
| 3 |   |   |   |   |   |   |
| 4 |   |   |   |   |   |   |
| 5 |   |   |   |   |   |   |
| 6 |   |   |   |   |   |   |

**(a)** What is the probability of getting an odd and an even number?

............................................................................................................................................................

............................................................................................................................................................ [1 mark]

**(b)** What is the probability of the product of the two numbers thrown being a square number?

............................................................................................................................................................

............................................................................................................................................................ [1 mark]

**(c)** What is the probability of the sum of the two numbers thrown being a factor of 24?

............................................................................................................................................................

............................................................................................................................................................ [1 mark]

199

**15.** This table shows data collected from a survey into the height of pupils and how far they could throw a shot putt.

| Height (cm) | 120 | 125 | 126 | 130 | 130 | 133 | 135 | 140 | 145 | 147 |
|---|---|---|---|---|---|---|---|---|---|---|
| Distance thrown (m) | 4.7 | 4.9 | 5 | 5.5 | 6 | 6.3 | 7 | 7.5 | 8 | 8.7 |

(a) Draw a scatter graph comparing heights with distance thrown. [3 marks]

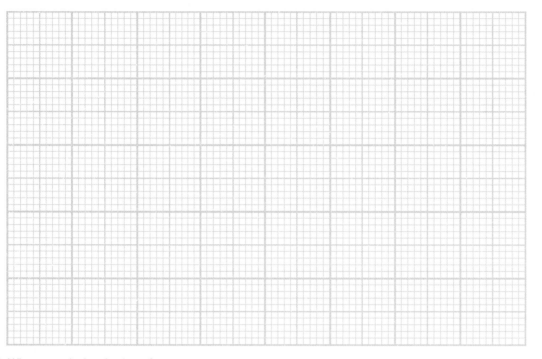

(b) What correlation is there?
What does this tell you about the connection between height and throwing distance?

_____

_____ [2 marks]

(c) Draw a line of best fit and estimate the distance thrown by a pupil of height 150cm and the height of a pupil who can throw 10m.

Estimated distance thrown = _____   Estimated height = _____ [3 marks]

**16.** A library manager recorded how often the computers were used each day in March.
A summary of her results is shown in this table.

| | Frequency of use |
|---|---|
| Minimum | 15 |
| Lower quartile | 24 |
| Median | 28 |
| Upper quartile | 34 |
| Maximum | 42 |

**(a)** Draw a box plot to illustrate these results.

[2 marks]

**(b)** This is the box plot illustrating July's results.

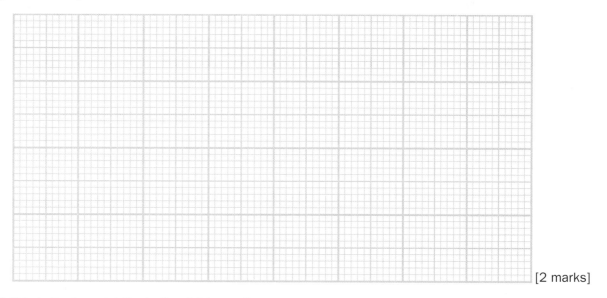

Write down two differences between the box plots and suggest reasons.

_____

_____ [2 marks]

**17.** Some fruit is delivered to a supermarket.

Out of a box of 50 apples, 5 are rotten and out of a box of 60 pears, 4 are not fit to be sold.

**(a)** What is the probability of choosing a pear that is fit to be sold?

_____

_____ [2 marks]

**(b)** What is the probability of picking out an apple that is not rotten and an unfit pear?

_____

_____ [2 marks]

**(c)** If three boxes of apples and three boxes of pears are delivered, estimate how many of each fruit would need to be thrown away.

_____

_____

_____

Estimated number of unfit apples = _____

Estimated number of unfit pears = _____ [3 marks]

# Further exam practice answers

## GCSE Paper 1

**1.** **(a)** **(i)** $70 = 2 \times 5 \times 7$
$84 = 2^2 \times 3 \times 7$ [2]
**(ii)** LCM of 70 and 84 $= 2^2 \times 3 \times 5 \times 7 = 420$
HCF of 70 and 84 $= 2 \times 7 = 14$ [2]

*Examiner's tip: LCM takes higher index and HCF takes lower index of factor.*

**(b)** $\frac{1}{5}$ of $5^6 = \frac{1}{5} \times 5^6 = 5^5$ [1]

**(c)** **(i)** $1\frac{3}{5} \div \frac{1}{6} = \frac{8}{5} \times \frac{6}{1}$
$= \frac{48}{5} = 9\frac{3}{5}$ [1]

**(ii)** $2\frac{3}{7} - 1\frac{2}{3} = \frac{17}{7} - \frac{5}{3}$
$= \frac{51 - 35}{21} = \frac{16}{21}$ [1]

**(d)** $100 \times 0.6\dot{1} = 61.616161$
$1 \times 0.6\dot{1} = 0.616161$
subtract: $99 \times 0.6\dot{1} = 61$
$\therefore 0.6\dot{1} = \frac{61}{99}$ [2]

**2.** **(a)** $\frac{43 \times 261}{197} \approx \frac{40 \times 300}{200} = 60$ [2]

**(b)** $(p^6)^{\frac{1}{2}} \times 5pq = p^3 \times 5pq$
$= 5p^4q$ [2]

**(c)** $3mn^2 - 9mn + 6m^2 = 3m(n^2 - 3n + 2m)$ [2]

*Examiner's tip: Use laws of indices for (b) and (c).*

**3.** **(a)** $0.000\,006\,130\,2 = 0.000\,006\,13$ to 3 s.f.
$= 6.13 \times 10^6$ [2]

**(b)** $52\,480\,000 = 52\,500\,000$ to 3 s.f.
$= 5.25 \times 10^7$ [2]

*Examiner's tip: Remember standard index form has a number multiplied by a power of 10.*

**4.** **(a)** $3x - 4 > 11 - 2x$
$3x + 2x > 11 + 4$
$5x > 15$
$x > 3$ [2]

*Examiner's tip: Collect like terms on same side of inequality.*

**(b)** Show your solution on this number line.

-1  0  1  2  3  4  5  6  7 [1]

**5.** **(a)** $3x^2 + 2y^3 = 3(2)^2 + 2(-1)^3$
$= 12 - 2 = 10$ [2]

*Examiner's tip: Remember signs when you are substituting values.*

**(b)** $\frac{1}{4} = 0.25$  $\frac{2}{5} = 0.4$  $\frac{3}{10} = 0.3$  $\frac{9}{20} = 0.45$  $\frac{3}{50} = 0.06$
$\frac{1}{2} = 0.5$ so 0.45 or $\frac{9}{20}$ is closest to $\frac{1}{2}$ [2]

**6.** **(a)** **(i)** Number of light
tile 1 : 4 squares
tile 2 : 8 squares
tile 3 : 12 squares
tile $n$ : $4n$ squares
**(ii)** Number of dark
tile 1 : 1 + 4 squares
tile 2 : 4 + 4 squares
tile 3 : 9 + 4 squares
tile $n$ : $n^2 + 4$ squares [4]

**(b)** Number of light $= 4 \times 8 = 32$
Number of dark $= 8^2 + 4 = 64 + 4 = 68$ [2]

**(c)** 30mm $= 30 \div 10 = 3$cm
number of small squares on tile with side measuring
21cm $= 21 \div 3 = 7$.
tile 5 has a side with 7 small squares [2]

**(d)** Ratio of measurements of tile sides
$= 9 : 12 : 15 : 18 : 21 = 3 : 4 : 5 : 6 : 7$
If tile 1 costs £1.50,
then tile 5 costs $\frac{7}{3} \times £1.50 = £3.50$ [3]

*Examiner's tip: Simplify ratio to its lowest terms.*

**7.** **(a)** $\frac{2m(6 + p)}{5 - p} = 3$
$2m(6 + p) = 3(5 - p)$
$12m + 2mp = 15 - 3p$
$2mp + 3p = 15 - 12m$
$p(2m + 3) = 15 - 12m$
$p = \frac{15 - 12m}{2m + 3}$ [4]

*Examiner's tip: Work towards isolating the new subject of the formula.*

**(b)** $y^2 + 2y - 15 = 0$
$(y + 5)(y - 3) = 0$
$y = -5$ or $y = +3$ [4]

**8.** **(a)** From the graph: day 1 costs £15
3 days cost £30
after day 1 each day costs $£(30 - 15) \div 2 = £7.50$
formula for the cost of hire($H$) for $d$ days:
Cost ($H$) $= 15 + 7.5(d - 1)$ [1]

*Examiner's tip: Check the formula by substituting values.*

**(b)** Line starts at £15. This is the charge for day 1 and is paid before days are calculated. [1]

**(c)** Cost to hire the saw for 5 days = £45 [2]

**(d)** Mike gets $4\frac{1}{3}$ day's hire for £40 [2]

**9.** **(a)** AM $= 5 \times \sin 50° = 3.83$cm
AB $= 2 \times 3.83 = 7.66$cm [2]

**(b)** $\angle$ACB $= \frac{1}{2}$ angle subtended at centre of circle
$= \frac{1}{2} \times 100 = 50°$ [2]

*Examiner's tip: Learn all about angles in circles.*

**10.** If C, D, E and F were all on the circumference, they would form a cyclic quadrilateral. This means the opposite angles would be supplementary or add up to 180°
$\therefore$ It is not possible to draw a circle, so that C, D, E and F are all on the circumference.
$\angle$C + $\angle$E $= 130° + 47° \neq 180°$
$\angle$D + $\angle$F $= 139° + 51° \neq 180°$ [2]

*Examiner's tip: Learn all the circle theorems.*

**11.** **(a)**

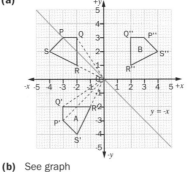

[2]

**(b)** See graph [2]

*Examiner's tip: Draw a line from the centre of rotation to a vertex to measure the angle.*

**(c)** See graph
Coordinates of the vertices of image B:
P (3, 3), Q (2, 3), R (2, 1), S (4, 2) [4]

**(d)** The column vector $\binom{-4}{0}$ would not move image B back to the original position. Image B is a reflection, in the $y$-axis, of the original shape. [1]

**12.** $3x + 2y = 6$

| $x$ | 0 | 1 | 2 |
|---|---|---|---|
| $y$ | 3 | 1.5 | 0 |

$x^2 - 5x + 6 = 0$
$x = 2.5$ gives minimum point

| $x$ | 0 | 1 | 2 | 2.5 | 3 | 4 | 5 |
|---|---|---|---|---|---|---|---|
| $y$ | 6 | 2 | 0 | -0.25 | 0 | 2 | 6 |

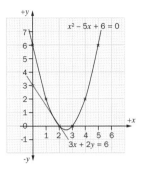

[6]

Points of intersection: $x = 2$; $y = 0$ and $x = 1.5$; $y = 0.75$ [2]

*Examiner's tip: The curve is symmetrical so continue beyond the calculated points.*

**13. (a)** Bearing of K from L = $80° + 180° = 260°$ [2]

**(b)** Scale of 1cm = 3km.

Your diagram should be drawn to scale.

Lengths must be 6cm, 3.3cm and 8.4cm.

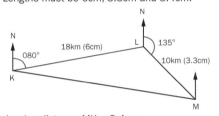

drawing distance MK = 8.4cm

actual direct distance MK = $8.4 × 3 = 25.2$km

bearing of M from K = $100°$ ($±2°$)

*Examiner's tip: Multiply by the scale to convert to the actual distance.*

**14. (a)** $\vec{CD} = \vec{CA} + \vec{AD} = \mathbf{-b} + \mathbf{c}$

$\vec{AM} = \frac{1}{2}\vec{AC} = \frac{\mathbf{b}}{2}$

$\vec{MN} = \vec{MA} + \vec{AN} = \frac{\mathbf{-b}}{2} + \frac{\mathbf{a}}{2} = \frac{(\mathbf{a} - \mathbf{b})}{2}$ [3]

**(b)** $\vec{AP} = \mathbf{c} + \frac{\mathbf{a}}{3}$ or $\mathbf{b} - \frac{2\mathbf{a}}{3}$

so $\mathbf{c} + \frac{\mathbf{a}}{3} = \mathbf{b} - \frac{2\mathbf{a}}{3}$

$\mathbf{a} = \mathbf{b} - \mathbf{c}$  ($\frac{\mathbf{a}}{3} + \frac{2\mathbf{a}}{3} = \mathbf{a}$)

$\vec{MP} = \vec{MA} + \vec{AD} + \vec{DP} = \frac{\mathbf{-b}}{2} + \mathbf{c} + \frac{\mathbf{a}}{3}$

$= \frac{\mathbf{-b}}{2} + \mathbf{c} + \frac{\mathbf{b}}{3} - \frac{\mathbf{c}}{3}$

$= \frac{2\mathbf{c}}{3} - \frac{\mathbf{b}}{6}$ (add fractions)

$\therefore \vec{MP} = \frac{4\mathbf{c} - \mathbf{b}}{6}$ [3]

*Examiner's tip: Remember direction when calculating vectors.*

**15. (a) (i)** $60°$  **(ii)** scale 1cm : 1m or 1 : 100 [1]

**(b)** Bisector of angle between walls

[2]

**(c)** Angle between the seedlings and a wall = $30°$ [1]

**(d)**

[2]

*Examiner's tip: Always leave the construction arcs on; you will gain marks for method.*

**16. (a)** $y = f(x) - 3$

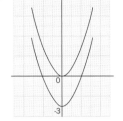

coordinates of the point where $y = f(x) - 3$

cuts the $y$-axis: $x = 0$; $y = -3$ [3]

**(b)** $y = -g(x)$

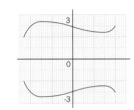

coordinates of the point where $y = -g(x)$

cuts the $y$-axis: $x = 0$; $y = -3$ [3]

*Examiner's tip: Learn transformations of functions.*

## GCSE Paper 2 Answers

**1. (a)** $5y + 6 = 18$

$5y = 18 - 6$

$5y = 12$

$y = \frac{12}{5} = 2\frac{2}{5}$ [2]

**(b)** $2(x - 1) = 3(5 - 2x)$

$2x - 2 = 15 - 6x$

$2x + 6x = 15 + 2$

$8x = 17$

$x = \frac{17}{8} = 2\frac{1}{8}$ [3]

**(c)** $z + 3 = \frac{4(2 - z)}{3}$

$3z + 9 = 8 - 4z$

$3z + 4z = 8 - 9$

$7z = -1$

$z = \frac{-1}{7}$ [3]

*Examiner's tip: Collect like terms on the same side in an equation.*

**2.**

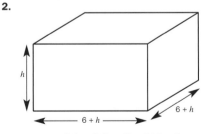

volume = $h(h + 6)(h + 6) = 600$cm³

$h^3 + 12h^2 + 36h = 600$cm³

| Trial | Result | Comment |
|---|---|---|
| 5 | 605 | Too large |
| 4.8 | 559.9 | Too small |
| 4.9 | 582 | Too small |
| 4.95 | 593.5 | Too small |

[3]

The exact value is between 4.95 and 5 so to 1 d.p. $h = 5.0$cm

*Examiner's tip: Try a sensible value first.*

**3. (a)** Actual cost of the car = $\frac{12\,599}{1.175}$ = £10722.55 [2]

**(b)** Total cost of the car would have been

1.15 × £10722.55 = £12330.93 [2]

**(c)** Percentage increase in total price after 1 January

= $\frac{12\,599 - 12\,330.93}{12\,330.93}$ × 100 = $\frac{268.07}{12\,330.93}$ × 100 = 2% [2]

*Examiner's tip: Remember to put increase over original cost.*

**4. (a)** AE = 5.5 × tan 56° = 8.15cm

$\tan 28° = \frac{AE}{CD + DE}$

CD + DE = $\frac{8.15}{\tan 28°}$ = 15.33cm

∴ CD = 15.33 − 5.5 = 9.83cm [3]

**(b)** AD = $\frac{5.5}{\cos 56°}$ = 9.84cm = CD

∴ Δ ACD is isosceles (2 equal sides) [2]

**(c)** Δ BCD and Δ ACE are similar triangles (3 equal angles)

This means the sides are in proportion:

$\frac{BD}{AE} = \frac{CD}{CE}$

BD = $\frac{8.15 \times 9.83}{15.33}$ = 5.23cm [2]

*Examiner's tip: Learn the trigonometric ratios.*

**5.** Arc of sector = $\frac{125}{360}$ × 2π × 10.5 = 22.9cm

radius of cone = $\frac{\text{arc}}{2\pi}$ = $\frac{22.9}{2 \times \pi}$ = 3.64cm

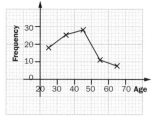

using Pythagoras: $h = \sqrt{10.5^2 - 3.65^2}$

vertical height of the cone ($h$) = $\sqrt{97}$ = 9.85cm correct to 3 s.f. [3]

*Examiner's tip: Remember that the sector is a fraction of a circle.*

**6. (a)** Volume of sand = 20 × 120 × 120 = 288000cm³

(1 000 000cm³ = 1m³) = 0.288m³ [2]

*Examiner's tip: Read the question carefully. Do not use full height; sand is only filled to 20cm.*

**(b)** (1m³ weighs 1.6t; 1t = 1000kg)

weight of sand = 0.288 × 1.6t

= 0.288 × 1600kg

= 460.8kg [2]

**(c)** Number of 15kg bags = 460.8 ÷ 15 = 30.72

= 31 to nearest bag

Cost at CostOK = (31 × £2.99) + £5 delivery charge

= £97.69

Number of 25kg bags = 460.8 ÷ 25 = 18.43

= 19 to nearest bag

Cost at BuyRite = (19 × £5.99) = £113.81

CostOK gives better value. [3]

**7. (a)** Volume of frustum = $\frac{1}{3}\pi R^2 H - \frac{1}{3}\pi r^2 h$

= $\frac{1}{3}\pi (10)^2 \times 12 - \frac{1}{3}\pi (5)^2 \times 6$

= $\frac{2}{3}\pi (200 - 25)$

= 2π × 175 = 350π cm³ [3]

**(b)** Volume of other cone = $\frac{1}{3}\pi r^2 \times 30$ = 350π

$r^2 = \frac{350}{10}$ (cancel by 10π)

$r = \sqrt{35}$ = 5.92 to 3 s.f.

curved surface area of cone

= πrl (l = slant height)

Using Pythagoras to find slant height: $l^2 = r^2 + h^2$

$l^2 = 5.92^2 + 30^2$

$l = 35 + 900 = 935$

= $\sqrt{935}$ = 30.6cm

curved surface area of cone = π × 5.92 × 30.6 = 569cm²

area of circular base = $\pi r^2$ = π × 35 = 109.96 = 110cm²

total surface area of cone = 569 + 110 = 679cm² [3]

*Examiner's tip: Check the question to see if the total surface area is required or just the curved surface area.*

**8. (a)** One lap of the track = (2 × 102) + (π × 60)

= 204 + 188.5 = 392.5m [2]

**(b)** Total running track = 6 × 1.25 = 7.5m [1]

**(c)** Diameter of ends that Sophie runs = 60m + 2 widths of track = 60 + 15 = 75

difference in runs = [204 + (π × 75)] − 392.5

= 439.6 − 392.5 = 47.1m [2]

**(d)** Area of turf = (102 × 60) + π(30)²

= 6120 + 2827 = 8947m²

= 9000m² to the nearest 1000m² [3]

*Examiner's tip: The two semicircles make a whole circle.*

**9. (a)** Total number of people = 90

median (middle) is the 45th/46th value which is in 40 ≤ x < 50 group

modal class is 40 ≤ x < 50 as frequency is highest [2]

**(b)** Midpoints of groups: 25, 35, 45, 55, 65

[4]

*Examiner's tip: Remember to plot at the midpoints of groups for a frequency polygon.*

**(c)** Number of people from each group in sample

= $\frac{\text{no. in group}}{90}$ × 30

= $\frac{\text{no. in group}}{3}$

so 20 ≤ x < 30 group = 6 people

30 ≤ x < 40 group = 8 people

40 ≤ x < 50 group = 9 people

50 ≤ x < 60 group = 4 people

60 ≤ x < 70 group = 3 people

(Part of a person is impossible) [2]

**10. (a)**

| Time (t hours) | Frequency (f) | Cumulative frequency |
|---|---|---|
| 8 ≤ t < 10 | 5 | 5 |
| 10 ≤ t < 12 | 10 | 15 |
| 12 ≤ t < 14 | 18 | 33 |
| 14 ≤ t < 16 | 13 | 46 |
| 16 ≤ t < 18 | 4 | 50 |

[2]

**(b)**

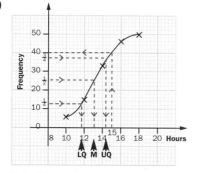

[3]

*Examiner's tip: It is useful to use a smooth curve for this graph.*

**(c)** Median = 13 hours (±0.5)
Inter-quartile range = UQ − LQ = 14.4 − 11.6
= 2.8 hours (±0.5) [2]

**(d)** 10 batteries last more than 15 hours.
Percentage of batteries that last more than 15 hours
$= \frac{10}{50} \times 100 = 20\%$ [2]

**11. (a)**

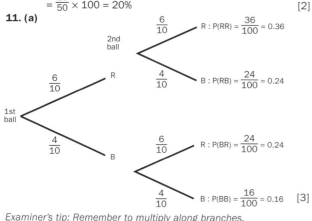

*Examiner's tip: Remember to multiply along branches.*

**(b)** $P(BB) = \frac{16}{100} = \frac{4}{25}$ (0.16) [1]

**12. (a)** Volume of the brick $= \frac{60}{360} \times \pi(3)^2 \times 9$
$= 42.4\text{cm}^3$ [3]

**(b)** Number of bricks forming a cylinder = 360 ÷ 60 = 6 [1]

**(c)** Circumference of base of cylinder $= 2 \times \pi \times 3 = 6\pi$
area of red on the cylinder $= 6\pi \times 9 = 54\pi = 169.6\text{cm}^2$ [2]

*Examiner's tip: Read the question carefully – only curved surface is needed.*

**13.** ∠ABC = 180° − 53° = 127° (interior angle of parallel lines add
up to 180°)
∠BCD = 180° − 48° = 132° (interior angle of parallel lines add
up to 180°)
∴ ∠DCE = 132° − 72° = 60°
∴ ∠BEC = 60° (alternate angles)
∠AEB = 48° (corresponding angles)
∴ ∠ABE = 180° − (53° + 48°) (angle sum of triangle = 180°)
= 180° − 101° = 79° [3]

*Examiner's tip: Learn all the facts about angles and parallel lines.*

**14.**

| | 1 | 2 | 3 | 4 | 5 | 6 |
|---|---|---|---|---|---|---|
| **1** | 1/1 | 1/2 | 1/3 | 1/4 | 1/5 | 1/6 |
| **2** | 2/1 | 2/2 | 2/3 | 2/4 | 2/5 | 2/6 |
| **3** | 3/1 | 3/2 | 3/3 | 3/4 | 3/5 | 3/6 |
| **4** | 4/1 | 4/2 | 4/3 | 4/4 | 4/5 | 4/6 |
| **5** | 5/1 | 5/2 | 5/3 | 5/4 | 5/5 | 5/6 |
| **6** | 6/1 | 6/2 | 6/3 | 6/4 | 6/5 | 6/6 |

**(a)** P(odd and even) $= \frac{18}{36} = \frac{1}{2}$ [1]

**(b)** P(product of two numbers being square number)
$= \frac{8}{36} = \frac{2}{9}$ [1]

**(c)** P(sum of two numbers being factor of 24) $= \frac{17}{36}$ [1]

*Examiner's tip: Always work methodically when filling out sample space diagrams, so no outcome is missed.*

**15. (a)** [3]

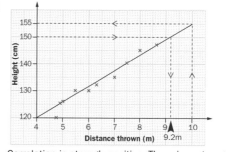

**(b)** Correlation is strongly positive. There is a strong
connection between height and throwing distance. [2]

**(c)** Estimated distance thrown = 9.2m
Estimated height = 155cm [3]

*Examiner's tip: Always ask yourself if your answer is sensible.*

**16. (a)**

[2]

**(b)** March range = 42 − 15 = 27; July range = 35 − 10 = 25
March IQ range = 34 − 24 = 10; July IQ range = 30 − 23 = 7
More use of computers in March than July.
Summer weather and activities better than in March. [2]

*Examiner's tip: As long as your reasons are sensible and relevant,
you will score marks.*

**17. (a)** P(pear fit to be sold) $= \frac{56}{60} = \frac{14}{15}$ or 0.93 [2]

**(b)** P(fit apple and unfit pear) $= \frac{45}{50} \times \frac{4}{60} = \frac{180}{3000} = \frac{3}{50}$ or 0.06 [2]

**(c)** Three boxes of apples = 150 apples
Three boxes of pears = 180 pears
Estimated number of unfit apples $= \frac{5}{50} \times 150 = 15$
Estimated number of unfit pears $= \frac{4}{60} \times 180 = 12$ [3]

*Examiner's tip: Use probability AND rule in part (b)*

# Notes

# Index

# Index